The Alewives' Tale

The Alewives' Tale

THE LIFE HISTORY AND ECOLOGY OF RIVER HERRING IN THE NORTHEAST

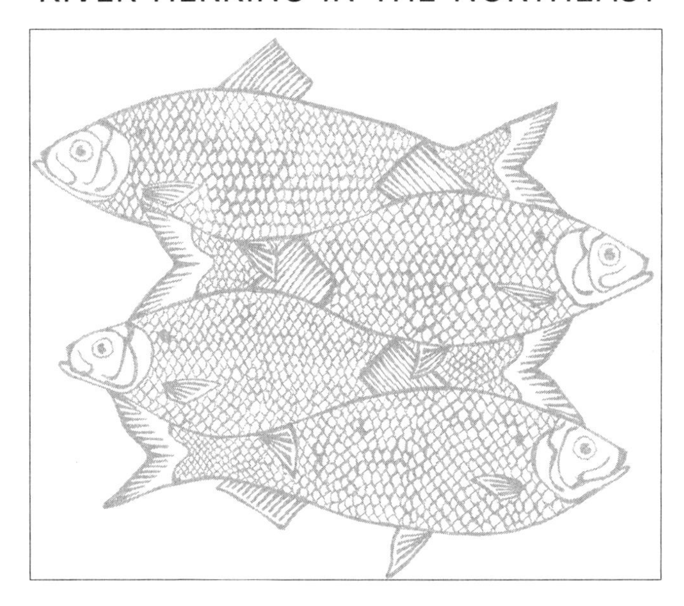

BARBARA BRENNESSEL

UNIVERSITY OF MASSACHUSETTS PRESS
Amherst and Boston

Printed in the United States of America

ISBN 978-1-62534-105-1 (pbk.); 104-0 (hardcover)

Designed by Sally Nichols
Set in Goudy Oldstyle
Printed and bound by Sheridan Books, Inc.

Library of Congress Cataloging-in-Publication Data

Brennessel, Barbara.
The alewives' tale : the life history and ecology of river herring in the Northeast / Barbara Brennessel.
pages cm.
Includes bibliographical references and index.
ISBN 978-1-62534-105-1 (pbk. : alk. paper) – ISBN 978-1-62534-104-4 (hardcover : alk. paper)
1. Alewife–Northeastern States. 2. Blueback herring–Northeastern States. 3. Fishes–Conservation–Northeastern States. 4. Stream conservation–Northeastern States. I. Title.
QL638.C64B74 2014
597'.45–dc23
2014019798

British Library Cataloguing-in-Publication Data
A catalogue record for this book is available from the British Library.

for all my "Friends of Herring River"

Contents

Acknowledgments

I am grateful to so many people for providing encouragement for this project and assisting with my interpretations of fish biology and fishery issues. John Duane and Abby Franklin were kind enough to read several chapters in their very rough stages and provide invaluable suggestions. Bob Prescott, John Riehl, Nancy Civetta, and my husband, Nick Picariello, made excellent suggestions for sections of the manuscript during the final editing stages.

I relied heavily on Stephen McCormack as I sorted through the cellular and molecular maze of osmoregulation, and on Ted Castro-Santos for his willingness to give me a tour of the Conte lab and his enthusiasm for all aspects of fish passage. Several scientists and fishery managers communicated with me by email; I am very grateful for their thoughtful and expansive comments. Thanks to Eric Palkovaks, for information about alewife genetics, and Steve Gephard from the Connecticut Department of Environmental Protection, Inland Fisheries Division, for information about river herring restoration projects in Connecticut.

Jason Stockwell reviewed the first two chapters of earlier versions of this manuscript and set me on a correct trajectory. He is a font of information and filled me in on alewife issues while I spent the day with him at Damariscotta

Mills. I am grateful for his otolith photo and his introducing me to Red's Eats. While he was at Gulf of Maine Research Institute in Portland, I visited Stockwell's lab, where Zack Whittener generously spent some time showing me how to dissect and preserve otoliths.

Amy Van Atten provided a thorough review of the fisheries observer program, complete with lab demonstrations and video clips of the actual protocols used by the observers. Greg Wells of the Northeast Fisheries Division of PEW Charitable Trust provided timely updates to fishery management plans.

Some local herring heroes were immensely helpful; these include Jeff Hughes, Wellfleet herring warden, and Frank Borek, alewife and shellfish warden in Brewster, and his wife, Miriam. Art Benner, of Alewives Anonymous, provided information about the dams and runs in Marion and Mattapoisett. James Gurney, Marion herring inspector, filled in some of the herring industry history. One of my first forays to an Alaskan steeppass fish ladder was made possible by Brian Eltz and Denise Savageau of the Greenwich, Connecticut, Conservation Commission, and Susan Baker, volunteer extraordinaire.

The folks at the Old Colony Historical Society in Taunton, Massachusetts: Jane Hennedy, director, Andrew Boisvert, archivist and library manager, and Christie Jackson, curator, were patient with my questions and helpful beyond words. Elizabeth Bernier was my go-to person for approval of images and poetry. Likewise, I am thankful for the assistance of staff at the Bourne Historical Society, Gioia Dimock, archivist, and Thelma Loring, Historical Commission curator, Mike Maddigan, vice chairman of the Middleborough Historical Commission, and Ronnie Peabody, of the Maine Coast Sardine History Museum. In gathering information about river herring, I had many questions. I appreciate the returned phone calls and emails from many Department of Natural Resources biologists.

Kathy Rogers, of Wheaton College, assisted in many logistical aspects relating to this book project. I always appreciate her thoroughness and ability to tackle any tasks with cheer and efficiency. My daughter, Marisa Picariello, contributed drawings. I am always indebted to her for helping me out with the artistic aspects of my projects. Also thanks to the rest of my family, who have learned that any road trip, from spring through fall, will include a stop at a herring run.

Brian Halley at University of Massachusetts Press took an interest in this project and shepherded it through its early rough stages. I am indebted to him for his support and guidance. It was a pleasure to work with Katherine Scheuer and Mary Bellino. I was amazed by their editorial attentiveness and efficiency.

We all should be appreciative of the efforts of the count volunteers, whose dedication has helped to erase the long neglect of these forage fish. They provide data regarding the run sizes, which highlight the overall plight of the river herring. I encourage everyone to become a volunteer. I guarantee that you will meet new and interesting friends and have another way to share in the anticipation and excitement of springtime in the Northeast.

Introduction

I was first introduced to river herring when I led my family, city slickers from Brooklyn, New York, on a hike within the Cape Cod National Seashore on Memorial Day weekend, 1980. One of our stops was the Sluiceway, a Wellfleet landmark consisting of a narrow and shallow stretch of water. The Sluice, believed to be the engineering feat of native people or early settlers to the region, connects two kettle holes, Higgins Pond and Gull Pond, and on the day we arrived it was teeming with small silver fish that successfully made their way across the small channel in water that was only a few inches deep. We watched in amazement. What determination! If one of the fish failed on the first attempt, it would build up some steam and try again. We lingered long enough to witness several hundred of them make the passage into Gull Pond.

I learned that the fish were alewives and they could be expected at the Sluice in spring during their annual run, and indeed I saw them every year in May. Then they seemed to disappear, and years might go by before another sighting. I attributed this to poor timing on my part and didn't give it much thought until about twenty-five years later when new signage appeared, which warned about mercury and indicated that fish caught at the ponds should not be eaten. This signage was followed in 2006 by signs prohibiting harvest of river herring, which at first I thought might also be due to the presence of mercury. These signs piqued my interest because I was not aware that anyone would want to catch and eat these particular small fish; they didn't look very tasty to me. However, after speaking to some striped bass fishermen friends, I heard their complaints about the ban and the loss of a source of excellent bait, and I subsequently started to pay more attention to the role and status of these ecologically important fish. After all, the kettle ponds are part of a wetland complex that that also includes a pond called "Herring Pond," and the ponds drain into a river, named the "Herring River," so . . . herring must have been abundant. Where are they now?

I learned more about river herring as Wellfleet began the exceedingly long process of restoring the Herring River with the goal of at least partial restoration of tidal flow and conversion of a degraded freshwater wetland back into a salt marsh. I became a herring count volunteer and board member of Friends of Herring River, an organization formed to promote and advocate for the restoration project. I started to attend many fisheries-related talks and events, only to find out that there appears to be much confusion, disagreement, and name-calling about who or what might be to blame for the declines seen in local alewife runs. Hoping to sort through the issue and get to the bottom of things, I reasoned that the truth might be distilled by examining the scientific findings . . . easier said than done. Scientists have been hampered from producing generally applicable sets of data for a number of reasons, including lack of interest by funding agencies, the expansive freshwater-marine geographical range of the species, the conflicting mission of agencies that have jurisdiction over management, the variability of existing data, and the fact that what was observed just ten years ago does not reflect the situation today.

I had already written about declines in other species: a reptile, the diamondback terrapin (*Malaclemys terrapin*), and a mollusc, the eastern oyster (*Crassostrea virginica*). It was not a surprise to learn that the reasons for the declines in these native species are multifaceted, interrelated, and complex . . . for example, overharvest, poor water quality, disease, predation, environmental degradation, regulatory agencies dropping the ball; and that the path toward conservation and restoration of these organisms and their historic habitat is complex. Although I started out studying and writing about particular species, I found that they could not be studied in a vacuum; an ecosystem approach was needed.

In considering a project on river herring, I found it helpful that I represent the neutral ground of interested parties who are concerned about river herring: neither a fisheries manager nor a commercial or recreational fisherman; I don't depend on fish for my livelihood or for enjoyment in my leisure time. I was drawn to the project because river herring are underdogs: they are not particularly charismatic, like the diamondback terrapin, and they are not gourmet items, like raw oysters. They are fighters: battling upstream each spring for a chance at reproducing, while benefiting other critters by maintaining large population sizes to serve their important role of forage fish-food for other fishes and animals to consume in great quantities. They may not be *The Most Important Fish in the Sea,* as H. Bruce Franklin describes menhaden, but they come pretty close.

A word about the somewhat inaccurate title of this book, *The Alewives' Tale:* the topic is river herring—a collective term for two distinct species, alewives and blueback herring. Because these species are closely related and difficult to distinguish upon cursory examination, in some regions they are lumped together under one name, "alewives," or, in the parlance of New Englanders, "herrin." So I offer apologies to blueback herring, whose name did not present a literary opportunity for incorporation into the title.

I have attempted to be consistent in my use of the plural for "fish." Biologists use "fish" when referring to multiples of the same species, but "fishes" when referring to two or more different species. When I refer to the plural for river herring, I simply use "fish," because often it is not apparent if the river herring are alewife or blueback herring or both.

I am happy to report that individuals and organizations are increasingly becoming interested in river herring. Although I have focused my project on the northeast United States, river herring are of similar concern in the Mid-Atlantic and southern states. These fish were an important part of our past, and they should be a part of our future. I write about them to interest others in their plight and summarize current information about them: issues leading to their decline as well as solutions that may lead to their recovery. Until their numbers increase and they reclaim their importance as part of our cultural heritage, you can find me each spring, rain or shine, at my post on the Herring River in Wellfleet, counting the alewives as they swim upstream.

The Alewives' Tale

CHAPTER 1
Counting on Herring

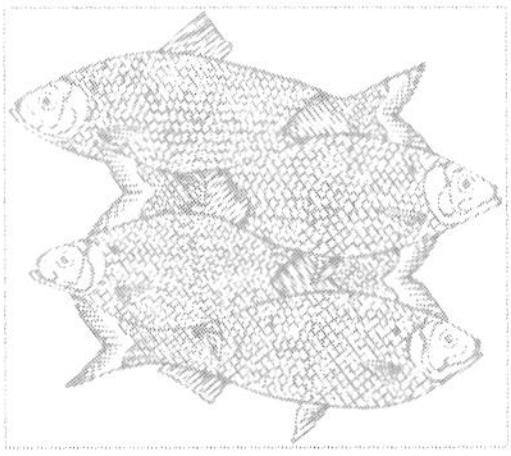

We watched, transfixed on the gently flowing water, for exactly ten minutes. With hand-held counters ready at our fingertips, we waited for signs of disturbance along the surface of the water, different from the whorls caused by the current, and for the silvery fish that would be making their way upstream to Herring Pond and beyond. My husband Nick and I were among the dozen or so individuals who had volunteered to count fish passing upstream in the Herring River in Wellfleet, Massachusetts, during the annual spring herring migration. The "herring count" was new to Wellfleet, but in 2009, we, along with volunteers in Chatham and Harwich, would be joining hundreds of others throughout New York and New England, who take short shifts to count the number of river herring that make the arduous journey to their spring spawning grounds.

As herring volunteers, we join a committed group of concerned individuals who count river herring (alewives as well as blueback herring) in many states along the Mid- and North Atlantic coasts, and our efforts are actually quite important. Our data provide a critical source of information to assess the number of river herring returning to our shores each year and, by extrapolation, the size of the entire river herring population. Fisheries experts agree that the picture looks very grim. In every location along the East Coast, river herring have been reduced to a small fraction of their historic numbers.

Early Harvests

Historical records suggest that river herring were abundant and also an important component in the diet of Native Americans. These oily fish, like other members of the family Clupeidae, are rich in Omega 3 polyunsaturated fatty acids such as EPA (eicosapentenoic acid) and DHA (docosahexaenoic acid) as well as Vitamin D, all known to be beneficial lipids. The Squamscotts and Wampanoags of New England relied on the predictability and timing of the annual appearance of the river herring and the relative ease of their capture to harvest a much valued dietary resource. There is archeological evidence of Native villages along the Nemasket River, which supports the largest run of alewives in Massachusetts, at sites where river herring were most easily harvested. Namasket means "place of fish" in the Wampanoag language. The river herring were salted, smoked, and dried to supplement the winter food supply of the native people. The Aquinnah Wampanoags of Martha's Vineyard still manage a herring run at Herring Creek, where the tribe has always fished.

Although not a staple, the herring were also part of the early colonial diet. Because herring could be salted, cured, and thus preserved, they were more popular as a food source in the era before refrigeration. Early European colonists in New England also used them for animal feed and as fertilizer. Eggs (known as roe), extracted from the females, were a delicacy and often found their way into omelets.

According to the writings of Captain John Smith, "But now experience hath taught them at New-Plimoth, that in April there is a fish much like a Herring that come up into the small Brookes to spawne, and where the water is not knee deepe, they will presse up through your hands, yea though you beat at them with Cudgels, and in such abundance as is incredible, which they take with that facility they manure their land with them . . ." (Smith 2007 [1630], 667). Smith's observations were confirmed in a letter to the Earl of Southampton, written on January 13, 1622, by John Pory, who was visiting Plymouth Colony on business. In describing the commerce of the colony, Pory wrote, "In April and May come another kind of fish which they call herring or old wives in infinite schools, into a small river running under the town, and so into a great pond or lake of a mile broad, where they cast their spawn, the water of the said river being in many places not above a half foot deep. Yea, when a heap of stones is reared up against them a foot high above the water, they leap and tumble over, and will not be beaten back with cudgels. . . . The inhabitants during the said two months take them up every day in hogsheads and what they eat not they manure the ground, burying two or three in each hill of corn" (Pory 1997 [1623], 7–8). Thus, river herring constituted an important subsistence fishery with no fish wasted: whatever was not eaten was used to fertilize crops.

It is believed that Squanto, the English-speaking Native befriended by the Pilgrims, helped the colonists survive at Plymouth Colony by instructing them how to use the river herring during spring planting to supplement the harsh New England soil. Squanto demonstrated for the colonists how to scoop up alewives at Town Brook and bury three fish with every five kernels of corn. The colonists were also advised by Squanto to guard their plantings so that animals would not dig up the fish before they decomposed. The colonists would thereafter "manure" their crops by burying alewives, as well as related fishes such as menhaden and shad when available, in mounds of soil where squash, beans, and corn were grown.

Brief Sketch of the Species

Although they are in the same biological family, Clupeidae, the river herring that are of concern in New England and the Mid-Atlantic are not the fish that are currently eaten raw, cured, and smoked and sold as smoked herring, kippers (spit and smoked), pickled herring, sardines, and other herring delicacies. These other popular herring are exclusively ocean dwelling and form large schools in both the North Atlantic (Atlantic herring, *Clupea harengus*) and the North Pacific (Pacific herring, *Clupea pallasi*). These *Clupea* species have been dietary staples in northern Europe, particularly in Scandinavian, German, Russian, and Eastern European Jewish cuisine, for thousands of years, and have thus supported large, commercially important fisheries, extensively described by Mike Smylie in his tribute *Herring: A History of the Silver Darlings* (Smylie 2004).

The herring of concern in Atlantic coastal rivers and streams consists of two similar species, collectively called river herring, related to *Clupea: Alosa pseudoharengus* (false herring), commonly known as the alewife, and *Alosa aestivalis*, commonly known as blueback herring. Because they are sometimes difficult to distinguish, they are grouped together for purposes of monitoring as well as for the commercial fishery. The only sure way to distinguish between the two look-alike fish is with a knife. In bluebacks, the color of the lining of the abdominal cavity, known as the peritoneum, is a sooty color, almost black, but in alewives, it has a paler, pearl gray or light pinkish tint. More subtle differences occur in the exterior morphology of the fish, which sometimes only an expert can distinguish: in the alewife, the dorsal area coloration, seen only in live or freshly caught fish, is gray-green to bronze-tone, while in the blueback it is a deep blue-green. In addition, the eyes of the alewife are larger than those of blueback herring, and overall, the blueback has a slightly smaller, more slender, elongated body while that of the alewife is deeper and more strongly compressed (fig. 1.1).

In some geographic regions, alewives and blueback herring are sympatric, an ecological term for species which inhabit overlapping habitat but do not interbreed; both species can be found in many of the same rivers and streams during spring migrations, but they have somewhat different habitat utilization and reproductive behaviors in these systems. Both are also found in overlapping locations in the marine environment along the Atlantic Coast. According to the National Marine Fisheries Service, the historic range of the alewife once extended into South Carolina; however, the current range is from Newfoundland to North Carolina. The range of the blueback herring is from Nova Scotia to northern Florida; thus, the blueback has a more southerly distribution and is more abundant in Chesapeake Bay and points south. Both species of fish were so ubiquitous that almost every town has a "Herring River" or "Herring Brook," and some have an "Alewife Brook." In Cambridge, Massachusetts, the northernmost station for the Massachusetts Bay Transportation Authority (MBTA) Red Line is the Alewife T stop, near Alewife Brook, a tributary of the Mystic River, where river herring were once abundant.

As a curious case of an introduced or invasive (nonnative) species, there are also large populations of alewives in the Finger Lakes in upstate New York and the Great Lakes between the United States and Canada. These landlocked populations are believed to be the result of invasion through the Welland Canal from the St. Lawrence Seaway into Lake Ontario and via the

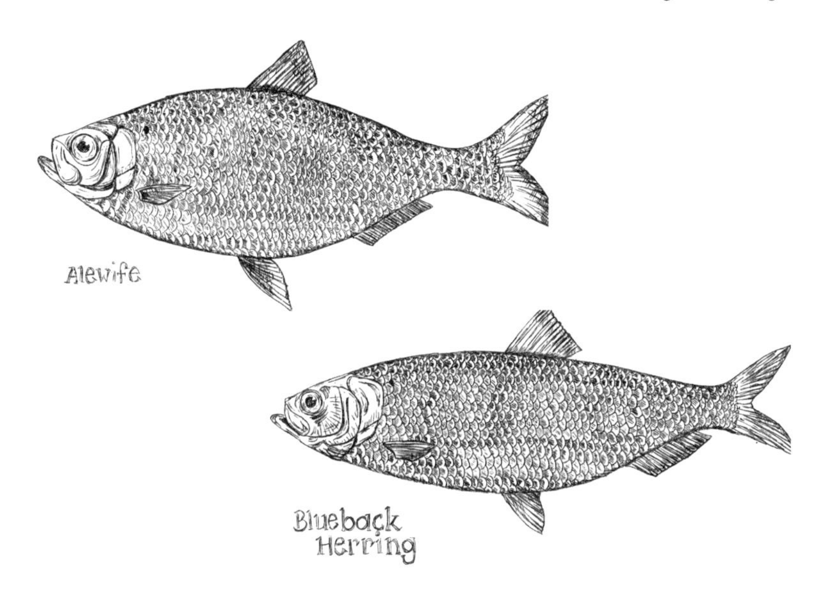

1.1 (top) Alewife: *Alosa pseudoharengus;* (bottom) Blueback herring: *Alosa aestivalis* (Marisa Picariello)

Erie Canal from the Hudson and Mohawk River systems. There are also landlocked alewives in other smaller northeastern lakes. These trapped alewives have adjusted to life in fresh water, and some of these populations are responsible for major changes in lake ecosystems.

Both species of river herring are small fish, silver in color with a black spot on each side behind the gills in mature fish. When fully mature, alewives once reached 36–38 centimeters (14–15 inches) in length while bluebacks were slightly smaller, only reaching a maximum of 32 centimeters (13 inches) (Haas-Castro 2006). In 2007, the Atlantic States Fishery Management Council published average sizes for adult alewives at about 25.4 centimeters (10 inches) long, weighing about 220–260 grams (8–9 ounces), and for bluebacks the average is now 27.9 centimeters (11 inches) in length and 217 grams (7 ounces) in weight, a significant reduction from historic sizes. River herring lack a lateral line, a sensory organ found in many fish, which detects changes in water pressure, vibrations, and movement. Such conditions are caused by changing depth and pressure waves, generated by currents due to movements of other nearby organisms. Thus, the lateral line system, absent in river herring, may allow fishes of other species to take evasive action from predators. The lower jaw of river herring protrudes slightly beyond the upper jaw, giving a bull dog-like appearance. They differ from *Clupea* ecologically in that they belong to a group of dozens of fishes that are classified as anadromous, i.e., they are marine fishes but they breed in fresh water and thus migrate between oceans and rivers or ponds. The term "sea run" is sometimes used to describe river herring and related anadromous species. The adjective "anadromous" and noun "anadromy" are from the Greek, meaning running upward. Similar migratory behavior is exhibited by such fishes as striped bass, rainbow smelt, American shad, Atlantic sturgeon, Atlantic salmon, and sea run brook trout, better known as charr or salters.

River herring were initially deemed significantly different from shad (*Alosa sapidissima*) and thus were formerly classified in the genus *Pomolobus.* They have been known by a variety of names, which are often local in origin and usage. Such monikers include: glut herring, summer herring, fresh water herring, blackbelly, kyak or kyack, wall-eyed, big-eyed, or blear-eyed herring (due to their large eyes), and sawbelly (due to the sharp serrations along the midline of their bellies formed by saw-toothed scales know as scutes). In Maine they were called cat-thrashers, and in Rhode Island they were referred to as buckies (Hay 1959). In French Canada, where they have given their name to several rivers, they are known as gaspereau. It is not clear how the name alewife originated. The old French and English names for the fish include allis, allizes, alouze, and aloose, and river herring were placed in a group of fish known as allis shad. One etymology stems from a supposed Native American name for bony fish, aloofe, which, in Connecticut, may have morphed to ellwhop, then ellwife, then alewife. This origin of the term seems unlikely if one considers that the Native term for alewife is *aumsuog,* according to the writings of Roger Williams. Other colonial-era authors refer to them as *umpeauges.* In 1901, as explanation that the term "alewife" did not derive from a Native name for the fish, it was pointed out in *Notes and Queries* (Gill 1901), a publication that answers readers' questions about English language, literature, and history, that *aloofe* may actually be a typographical error for *aloose,* after the old French word for shad, especially given that the letter *s* in earlier French texts could be written in a way to resemble the letter *f.* A more colorful origin of the name might be traced to taverns or alehouses where it was the custom to serve the fish, salted, as a snack to eat with the local ale which was brewed by the woman of the establishment, the so-called ale woman or ale wife. Perhaps the large abdomens of these portly women resembled the deep belly of the fish, as noted in 1674 when John Josselyn wrote, "the Alewife is like a Herrin, but has a bigger bellie, therefore called an alewife" (Josselyn 1988 [1674], 221).

Declining Stocks

Throughout the seventeenth and eighteenth centuries, the annual supply of river herring was an important source of protein in the diet (McKenzie 2010). However, by the 1800s and early 1900s, river herring were also used commercially for pet food, fertilizer, and bait for cod, haddock, pollock, and mackerel. Until late 1960, river herring were harvested exclusively near shore or in fresh water, where they once supported significant seasonal fisheries for commercial enterprises as well as recreational anglers. During the heyday of the fishery, World War I and a brief period afterwards, river herring were also a source of pearl essence, a substance more commonly seen as a lucrative by-product of the larger Atlantic herring fishery. Pearl essence is a translucent substance extracted from the silvery fish scales and used for production of paints, ceramics, costume jewelry, and cosmetics such as lipstick and nail polish. On Martha's Vineyard, Edgartown resident Ralph Bodman devised a way to coat glass beads with herring scales to make artificial pearls, called "Priscilla pearls," which were strung into necklaces.

Recreational harvesters of river herring use the fish as bait for bluefish and striped bass, and most recently river herring have been harvested commercially for crab and lobster bait. The U.S. commercial harvest peaked in the 1950s at over 34,500 metric tons before the very dramatic nationwide decline in the 1970s (Haas-Castro 2006; Limburg and Waldman 2009). Figure 1.2, created with data gathered by the National Marine Fisheries Service (NMFS) and reproduced in public documents such as the Interstate Fishery Management Plan for Shad and River Herring, depicts the sad and dramatic story in a graph of total commercial landings of river herring (both alewife and blueback herring) from 1950 to 2006.

The downward trajectory of the river herring populations is not a new or sudden phenomenon. The documented slow and steady decline of the fish had its beginnings during the time of European settlement when anyone could catch as many fish as they desired. Conflicts among fishermen and

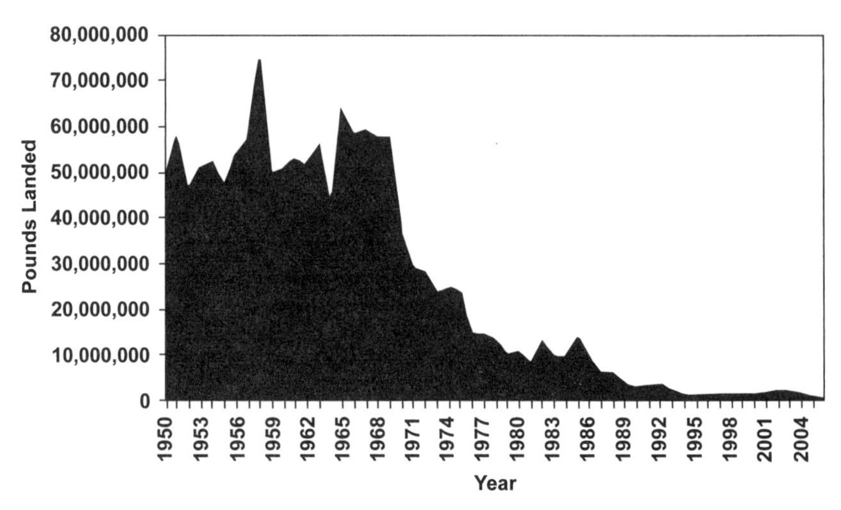

1.2. Commercial landings of river herring, 1950–2006 (NMFS)

between fishermen and early industrialists, who were building dams across prime fishing areas, resulted in a series of early colonial laws, designed to regulate and protect the fishery. Ineffective legislation and ineffective enforcement of the regulations resulted in fewer river herring arriving in spawning areas and a general decline in the populations. This trend has been followed by more abrupt declines in more recent times. As early as the mid-1800s, Canadian maritime province officials voiced concern about the declining river herring fishery. In nearby Maine, three days per week closures of the fishery have been in effect for many years. In Waldoboro, on the Medoomak River, a six-year moratorium on the taking of alewives was imposed in 1875 to allow stocks to rebound. During the 1970s, alewife harvests exceeded 3.4 million pounds annually, according to the Gulf of Maine Research Institute, making this fishery the most economically important anadromous fishery in the state, but, by 1988, only five hundred bushels were harvested. The U.S. Fish and Wildlife Service and Maine Department of Natural Resources report that alewives are no longer seen in Alewife Brook in southern Maine.

In Massachusetts, the alewife fishery was thoroughly evaluated in 1920 by Dr. William Belding in his *Report Upon the Alewife Fisheries of Massachusetts*. Belding used surveys of coastal streams, statistical studies of fishing methods, and investigations of the life history of alewives to assess the herring population and threats to the species and concluded that the condition of the fishery was "impoverished" (Belding 1920). Belding reported that many formerly productive runs had become garbage dumps and were filled with debris. Local Massachusetts economies no longer depended upon alewives, and few people seemed to care about the fish or their habitat. In 1900, millions of fish were seen in the Mattapoisett River near Buzzards Bay, but by 2000 the number had declined by at least one order of magnitude, and it was reduced to approximately five thousand in 2004. Similar trends have been observed in the Sippican and Bournedale Rivers, which also feed into Buzzards Bay.

Near Boston, the river herring fishery all but disappeared by 1920, when the fish seemed to vanish from the Merrimack, Mystic, Charles, and Neponset Rivers. Just north of Boston, a herring count was initiated on the Parker River in 1997 and was extended to include the Essex, Ipswich, and Saugus Rivers. By 2005, volunteers had recorded the lowest count on record in the Parker River: fewer than one hundred fish were seen, orders of magnitude lower than 1998 and 1999, in which the highest counts were observed. The annual counts suggested declines throughout the region and that the rivers were supporting only a fraction of their potential, based on the extent of suitable habitat (Purington, Doyle, and Stevenson 2003).

In Rhode Island, a similar story has unfolded. In 2000, 290,000 fish were counted at the Gilbert Stuart run (the largest in the state), but these numbers declined 90 percent by 2006. In March 2006, Rhode Island announced emergency regulations declaring a ban on the taking of river herring in both marine and fresh waters. The same scenario of precipitous decline has been documented in Connecticut, where the river herring fishery was once considered a significant industry. The Connecticut River has supported runs of alewives and blueback herring for thousands of years. Alewife were found in the Connecticut River as far north as the Massachusetts boundary, while blueback herring were known to migrate into New Hampshire and Vermont. Several Connecticut herring runs, including those on the Connecticut River, have been monitored for long periods of time: the Mianus River in Greenwich, the Branford Supply Pond Dam, Bride Brook in East Lyme, and others. Whether it is the largest river, the Connecticut, or smaller coastal rivers and streams, the numbers tell of a serious and abrupt decline in the numbers of fish everywhere.

Sparked by declining populations of river herring throughout their range, there is also a large volunteer herring count effort in New York where herring typically ran along 152 miles in sixty-five major Hudson River tributaries.

Due to the enormous expanse of the river herring migration route in New York, counting efforts are focused on a limited number of streams that drain directly into the Hudson.

Typical of other Mid-Atlantic states, Maryland has witnessed a dramatic decline of river herring since the mid-1970s. Prior to this time, there was extensive harvest of the fish, commercially as well as recreationally, in most rivers and tributaries that feed into Chesapeake Bay. "In 1931, over 25 million pounds of river herring were harvested, ranking them 2nd in quantity and 5th in value of all Chesapeake finfish, and 1st in quantity and 4th in value of all finfish landed in Maryland" (Maryland DNR). In Maryland, the commercial and recreational river herring fishery is currently closed from June 6 to December 31 of each year.

As one travels south along the Atlantic Coast, the story develops in a similar fashion. Hightower, Wicker, and Endres (1996) speculated that high harvest levels over short periods of time, coupled with offshore catches from vessels from foreign countries, were responsible for the decline of river herring landings in North Carolina in the mid-1990s to 5 percent of the amount of reported landings in the period from 1880 to 1970. A 2005 stock assessment in Albemarle Sound, N.C., indicated high mortality rates, decreased recruitment (fewer younger fish), and reduced numbers of spawning adults—all signs of a serious continuing decline.

Since 2006, the National Marine Fisheries Service of the U.S. government has classified river herring as a "Species of Concern." In May 2009, the Atlantic States Marine Fisheries Commission amended their Interstate Management Plan for Shad and River Herring (Amendment 2) to prohibit all commercial and recreational harvest of these fishes as of January 1, 2012, unless individual states can demonstrate that they have a management plans for sustainable harvest. The state of Maine regulates the river herring fishery in such a manner that individual cities and towns have local control. In 2009, LD 151, an Act to Amend the Alewife Fishery, was passed by the Maine legislature. The Act allows the Maine Department of Marine Resources to lease fishing rights in any river or stream that is not under municipal control. In Maine, the fishery is an important seasonal source of revenue for about three hundred individuals, mostly from rural areas. There is no "directed" offshore fishery for river herring, but in terms of inland harvest, Massachusetts (fig. 1.3), Rhode Island, Connecticut, and North Carolina offered legal protection for the fish and banned all harvest, at least on a temporary basis.

1.3. Sign in Massachusetts at the Herring River in Wellfleet (photo: Barbara Brennessel)

Many factors have contributed to the dramatic declines in river herring populations throughout the twentieth century. In many locations, impediments to upstream migration, such as dams, block access of river herring to historic spawning grounds. Loss of river herring habitat can also be traced to degradation of estuaries, rivers, and streams by diversion of water, siltation, and general loss of wetland areas. Another threat, overharvest, has surfaced off the coast. Fisheries documentation indicates that from 1966 to 1977 and 1984 to 1989 river herring were commercially harvested in large quantities off-shore. These fishing efforts were conducted by Mid-Atlantic fishing vessels from countries including Cuba, Bulgaria, Germany, the Netherlands, Poland, Spain, and others (NOAA Fisheries 2009). It is not clear how these harvest levels contributed to the documented declines of the fish as monitored during their inland migration. River herring are also caught as bycatch in other directed Atlantic fisheries. Such bycatch, when added to the losses of river herring by direct harvest and other factors, can make a significant dent in the population.

Why should we be concerned about a humble fish that lacks the charisma of other inhabitants of Earth's waters such as whales, sea turtles, salmon, and dolphins, whose populations are also in decline? The answer to this question is rooted in the ecological web, where a small creature can make a big difference in the scheme of things. The large historic numbers of river herring bound together the Atlantic Ocean with our East Coast rivers, streams, ponds, and lakes, creating a two-way flow of nutrients between freshwater and marine ecosystems. Their commercial and recreational value has all but disappeared. However, as with other small schooling fishes, their ecological role is not glamorous but it is very critical; these small creatures are an important forage base, providing a source of protein and lipid for other fishes, birds and mammals, i.e., predatory species which we hold in high esteem: striped bass, bluefish, tuna, cod, halibut, ospreys, otters, great blue herons, porpoises, dolphins, and many others. Young river herring are consumed by freshwater species such as bass, trout, perch, pickerel, and pike. Furthermore, the story of river herring is intimately linked with the story of the decline of our rivers and watersheds, which have been altered in many ways, intentionally and unintentionally, by human activities.

The Value of the Count

River herring that migrate to specific freshwater breeding areas are often considered to be unique stocks, but it is difficult to obtain a reliable stock assessment, primarily because these fish move between diverse habitats: ponds, lakes, rivers, streams, and the ocean. Historic harvest records provided useful data about population abundance in the days when there was a commercial harvest. There are other methods for counting herring to determine population size that do not rely heavily on harvest data or estimates based upon statistical sampling methods with the efforts of volunteers. For example, the Essex Dam on the Merrimack River in Lawrence, Massachusetts, has a fish lift, which is basically a herring elevator. As fish are lifted in buckets to the top of the dam, each bucketful can be directly counted.

If migrating herring pass through narrow sections of streams, or if they pass through a man-made structure such as a fish ladder or fish lift, they can be counted with the aid of an electronic counter. The counters may also rely on direct visualization via video cameras. In New England locations, such

1.4. Mianus River and Fishway (photo: Barbara Brennessel)

as Bournedale, near the Cape Cod Canal, the Nemasket River in southeast Massachusetts, Bride Brook near Long Island Sound, and on the Mianus River in Greenwich, Connecticut, electronic fish counters are in place; so as each fish passes by, a sensor automatically adds it to the tally.

The Greenwich Conservation Commission hosts an open house each spring at the Mianus fishway (fig. 1.4), which provides a firsthand view of a fish counter in operation. The Mianus Pond Dam was constructed in 1926 to impound water that was then diverted to cool a nearby power plant. The negative repercussions on anadromous fish migration followed very shortly. In 1993, as a result of a joint effort among local groups, the Connecticut Department of Environmental Protection, and the Greenwich Conservation Commission, funds became available to construct a fishway using a design known as an Alaskan steeppass. For good measure, an eelpass, a simple, rope-like device that is placed directly on the face of the dam, was also installed. My daughter, who was working for Soundwaters, an environmental education and stewardship organization, alerted me to the 2010 open house event and instructed me to go to the River House, a Greenwich adult day care center, adjacent to the fishway. It turns out that the day care center was once the cold water pumping station for the power plant.

I arrived at the center and met Denise Savageau, Greenwich Conservation Director, as well as ultra-enthusiastic volunteer Susan Baker, who was equipped with a long-handled scoop net filled with alewives, which she was showing to a visiting family. Baker held up one specimen and explained how the dark dorsal coloration helps the fish escape predators. The Mianus fishway supports one of the largest runs of river herring in the state of Connecticut. Brian Eltz, Conservation Assistant for the Town of Greenwich, pointed out the fish counter and explained how it works. Eltz used the feedback from an underwater video camera to elucidate the operation of the counter. The fish pass through eight stacked PVC pipes, each with its own electronic counter. The diameter of the pipe accommodates one fish at a time and as the fish pass through they break a current across the pipe, which registers on the counter. This system provides reliable information about the number of fish that are able to circumvent the dam. The fishway also has a webcam so that even if folks can't make it to the open house, they can log in and watch the action.

Unfortunately, sophisticated fish counters such as the one at the Mianus fishway are expensive and sometimes difficult to employ, especially in some of the smaller coastal streams. In order to effectively manage the herring fishery, the Massachusetts Division of Marine Fisheries (MA DMF) depends heavily on the volunteer herring count effort to supplement the data obtained with electronic counters and thus assess the status of herring abundance, or lack of abundance, throughout the state. The volunteer herring count on Cape Cod, one of many such programs in Massachusetts, is orchestrated by JoAnn Muramoto, working for the Association to Preserve Cape Cod (APCC) as regional coordinator of the Massachusetts Bays Program. With the help of local herring wardens, Muramoto trains volunteers and supplies them with low-tech tools: a hand-held counter, a thermometer, pencil and paper, all contained in a small zip-lock plastic bag. The training and protocols for volunteers are based on statistical concepts, described by Nelson (2006), that would allow an estimation of the number of herring, i.e., the run size, based on visually counting herring during ten-minute intervals. Volunteers sign up for random ten-minute shifts during three time periods: 7–11 a.m., 11–3 p.m. and 3–7 p.m. The count usually begins when water temperature reaches 10°C (50°F), and ends when fish can no longer be observed traveling upstream. In Wellfleet, this results in a sampling effort that is usually spread out over two months: April and May.

At a 2011 workshop, fisheries biologists, herring wardens, and volunteer count coordinators from Massachusetts met at Jones River Landing in Kingston to learn how to standardize methods and coordinate reporting of volunteer counts. Sara Grady, of the North and South Rivers Watershed Association, designed a database that would allow counts to be entered by coordinators and interface with a database that has been created by the MA DMF. Throughout the state, herring volunteers also measure water temperature and make observations about rainfall, air temperature, and other conditions. The data are important, even if there are no fish observed because the absence of fish is also critical information. The volunteer herring count is part of a state and regional effort to assess the status of the herring population and identify the causes for the recent precipitous decline. The river herring fishery, commercial as well as recreational, has been closed in Massachusetts since 2006.

During my spring travels I also met dedicated herring volunteers at other monitoring sites such as Hinkley's Pond in Harwich and Stony Brook in Brewster, as well as many folks who were at strategic viewing spots just to see the show. The sites are commonly referred to as herring runs or, simply, runs. A visit to the local herring run, associated with the coming of spring, is a ritual for many families. "The herring are running!" is the news that draws visitors to fishways, those sections of rivers and streams that are accessible to the public and where the herring are readily visible. The locations are often narrow sections of a waterway that funnel the fish schools into narrow silvery ribbons, but these viewing locations can also be located at man-made structures such as fish ladders adjacent to dams. These ladders and other such devices have been installed, some since colonial times, to offer the migrating fish a manageable upstream detour to their spawning grounds.

Fishways

Some of the most spectacular fishways in New England are without doubt the ones that are part of the Connecticut River watershed. Here, visitors can view determined, struggling fish, including some river herring, though large glass windows. A 1872 Supreme Court ruling mandated the construction of a mechanism for fish passage at the Holyoke Dam so that upstream residents would not be denied their fishing rights. In the mid-1900s, FirstLight Power Resources and Holyoke Gas and Electric Company were required to install a fish lift, the first of its kind in New England, at the Holyoke Dam to allow the passage of fishes to upriver spawning sites. The directives were aimed primarily at shad, later at salmon. The Robert E. Barrett Fishway, constructed in 1955, consists of two fish lifts, one for the river itself and one for the power canal. A loud alarm, sounding at pre-set intervals, is the signal that another batch of fishes are scheduled for the elevator. The fishes, milling near a man-made current that attracts them into the apparatus, are herded into the lifts by moving gates and then elevated to the level of the river upstream of the dam. As the fishes exit one of the lifts, they pass by a viewing window, open to the public on certain days in May and June. When I visited on a hot spring day in 2010, most of the passing fishes were shad; indeed a shad derby was occurring near the dam on the same day. There were also some lampreys and a solitary blueback herring in the viewing window. It was disappointing to see a schooling fish like the blueback, without others of its kind. Only forty blueback herring were seen at the dam in 2009, while during my visit in 2010, the tally for the year was fifty-eight. Alewives, on the other hand, have never made it up as far as Holyoke in recent history. I also stopped at the Rainbow Fishway, a bit farther south along the river, but it was closed for viewing owing to lack of personnel. However, I did see some dead shad being knocked down the fish ladder by the current. Apparently, the fishways on the Connecticut River were designed for salmon with the result that other fishes, smaller and less powerful than salmon, are sometimes fishway casualties.

Perhaps the best-known fishway in New England is the herring run at Stony Brook, in the town of Brewster on Cape Cod, made famous by John Hay in his lyrical and meditative classic, *The Run.* Hay's book, written in 1959, has become the river herring bible and is a must-read book for herring wardens, herring volunteers, and anyone else who has marveled at the life history of this fish and its long and treacherous journey from inland streams and ponds to the sea and back again. The first time I visited the Stony Brook Run was on a fine April day when my initial observation was of an audible nature: the loud shrieking of gulls. Indeed, the gulls are drawn to the runs, much as we are, but for different reasons. For us, the run is a glimpse of nature, a view into the life history of a familiar creature, a rite of spring, a sign of renewal. For the gulls, the run is a fast food stop. They easily pick off the fish as they collect in pools, or as they make their way up the stream, often powering their streamlined bodies up a foot of falling water, using their muscles to propel themselves upward to surmount a small boulder or a man-made step in a fish ladder, but sometimes they end up being swallowed whole by an awaiting gull (fig. 1.5). The sound of the gulls is not the only sign of the fish migration;

1.5. River herring gauntlet at the Stony Brook Run in Brewster (photo: Barbara Brennessel)

as the gulls perform their gluttonous ritual, they leave behind signs of their activities. Guano is everywhere. It coats the rocks lining the stream and whitewashes the roofs of nearby homes.

The feeding frenzy I observed at Stony Brook is not uncommon during the spring herring migration. A similar scenario was observed and reported by Tim Visel at Bonnet Shores, in Narragansett, Rhode Island. At his observation post at the mouth of a stream, Visel observed large striped bass, replenishing their reserves after migrating from southern waters, "thrashing along the surf" and the arrival of ospreys, needing energy-rich food for their young, swooping down from above. He felt sorry for the fish that had "overcome thousands of miles at sea to return to their birthplace only to run into this gauntlet of carnage!" (Visel 1988). Scavengers such as raccoons and skunks will also make their appearance during the migration, cleaning up the remains of the feast so that only piles of fish scales remain.

The Brewster run is not the longest or most commercially important run, but it is among the most scenic. The source of the run can be found in three main ponds, Upper Mill Pond, Lower Mill Pond, and Walkers Pond, and several smaller ones, all interconnected by narrow streams. The downstream path can be traced from the ponds, through the run and into a meandering stream known as Paine's Creek, before it empties into Cape Cod Bay. Between the ponds and the creek, at the falling headwaters where the run is sited, the town inhabitants placed a water mill. Today, the Stony Brook grist mill on Satucket Road, originally constructed in 1699, rebuilt in 1761, is still operational, but serves mainly as a tourist attraction. On one of my visits to Stony Brook, I met Herring (and Shellfish) Warden Frank Borek and his wife, Miriam, who visit the run during the herring migration and continue to volunteer at the mill during corn grinding season, when the corn is crushed for tourists and the fresh-ground cornmeal can be purchased. Frank is a retired teacher, apparent from his discussion with me in which he colorfully and patiently recounted his observations about the Stony Brook Run and the herring that pass through each year. He estimated that it takes each fish about one and a half hours to get from Cape Cod Bay to the Mill, a distance of about a mile and a half, a speed of about one mile per hour. Frank considers the run, the mill, and the surrounding area to be the "nicest free public park on Cape Cod." He reads John Hay's book once a year in its paperback edition and keeps his signed, first edition copy of *The Run* pristinely protected in plastic.

Making Sense of the Census

In November 2009, many of the Wellfleet herring volunteers gathered at the Wellfleet Elementary School for the Annual State of the Harbor Conference where JoAnn Muramoto was scheduled to give an update on the Cape Cod herring populations and, in particular, the first estimated number of herring that constituted the spring migration in the Herring River, Wellfleet. The overall news was disappointing in runs for which there were long-term data. The historic Stony Brook Run had declined dramatically compared to 2008. The official count was 233 herring, which gives a statistical estimate of a run size of about 11,000 fish. In Wellfleet there was no baseline for comparison, but volunteers were able to observe 1,664 fish, which extrapolated to an estimate of about 17,000. However, count volunteers reported that most of the fish passed the count location near dusk, just before the official 7 p.m. end of the daily observation period. Wellfleet Herring Warden Jeff Hughes has been a keen observer of the river since his boyhood days. He believes that the fish may be moving upriver under cover of darkness, a pattern that is also characteristic of Bride Brook and other runs in Connecticut. A night run would mean that the volunteer count results may be severely compromised. While the significance of the single 2009 count number cannot be assessed in a vacuum, at least there is a reference number for future comparisons.

Although the estimates of a herring run size may not be accurate, the number is a very important criterion for future planning and management. When faithful and diligent observers anecdotally report a decline in the run size, it can indicate a serious problem. However, to make more reliable comparisons, actual numbers, which can be evaluated from year to year and over the long term, are important for those entrusted with management of fisheries and other precious resources. Now Wellfleet has a number to work with, and those of us who are herring count volunteers will continue to hike to our observing posts because it will be even more important in succeeding years to make the count that will illustrate whatever trends may be occurring in the Herring River system.

To begin to understand the cause for the dramatic and somewhat sudden decline in the numbers of river herring, to address the causes of their imperilment, and to create meaningful solutions for restoring their populations, it is important to understand their life history and reproductive strategy. It is also crucial to understand the habitats that are significant for their survival—most important, the rivers and watersheds in which they spend a very brief, but extremely critical part of their lives.

CHAPTER 2
Return of the Natives

When I see the river banks
dusted with snow
I wonder how
the herring know
the time is right
for their perilous plight
the struggle upstream
perhaps at night. . . .

tomorrow may warm
the next day be cold
but nature plans well
preparations unfold
there's much to combat along the way
and I wonder how many made it today?

–From "April," by Ernestine Gray, a Native American and Bourne resident, who wrote this poem in the 1980s as she observed the declining numbers of herring in the run that had awed her from childhood.

The life of river herring is spent in three different types of ecosystem: the marine environment with salinity averaging 32 ppt (parts per thousand), estuarine systems that are brackish, with a mix of salt and fresh water, and freshwater systems, consisting of rivers and their tributaries, streams, and connecting lakes and ponds. An understanding of the life history of these anadromous fish and how it is intertwined with human settlement and activities offers the potential to shed some light on the current decline of their populations. In addition, such an understanding will be the foundation for solutions, which may help the fish populations to stabilize and perhaps rebound in the future.

Life at Sea

Adult river herring are schooling, sea-dwelling fish that inhabit waters near the Continental Shelf between spawning migrations. Here, they spend most of their lives. Although blueback herring have a more southerly distribution, there is much overlap in the oceanic ranges of the two species. The relative distribution of the species trends with ocean temperatures, but may also reflect the abundance and seasonal shifts of the river herrings' food supply.

Bottom trawling surveys and catch records over sixteen years, from 1963 to 1978, have shed some light on the oceanic movement of river herring. During winter, the pelagic schools cluster in northern latitudes between 40° and 43°, but by spring they can be found throughout the Continental Shelf between Nova Scotia and Cape Hatteras, North Carolina, with a great concentration at the Mid Atlantic Bight, that region of the Atlantic Coast between Massachusetts and North Carolina. During summer and fall, adults can be found north of 40° latitude, clustered in Nantucket Sound, Georges Bank, and the perimeter of the Gulf of Maine (Neves 1981; Pardue 1983). Due to recent dramatic stock declines, there has been an active effort to update these surveys and more finely characterize the seasonal oceanic movements of river herring in efforts to mitigate against the possibility of further declines due to bycatch, especially in the directed fishery for Atlantic herring. Using trawl surveys from 1948 to 2008 as well as catch data, the Atlantic States Marine Fisheries Commission is leading an effort to delineate the movement of alewives and blueback herring by producing maps which indicate the most likely areas where river herring are schooling at sea at different times of the year. These geographical and temporal maps can be used to identify seasonal hot spots so that possible future adjustments to the commercial fishery will be more likely to reduce river herring bycatch.

On the basis of harvest records (Neves 1981), the Central New England Fisheries Resource Office of the U.S. Fish and Wildlife Service estimates that alewives are more abundant at greater depths than those at which bluebacks concentrate: 56–110 meters (184–361 feet) for alewives compared to 27–55 meters (86–180 feet) for bluebacks. Neves speculated that the preferred depth ranges for each species may correlate with eye size. The alewife, with slightly larger eyes, may be better adapted to see at greater depths. Neves also questioned whether different dorsal coloration for each species may represent "a counterstrategy mechanism for reduced predation within the depth ranges most frequently occupied by each species" (Neves 1981, 483).

Seine hauls indicate that the schools may contain thousands of individuals. The two species sometimes school together and have also been observed mixing with other schooling fish such as Atlantic herring (*Clupea harengus*) and Atlantic menhaden (*Brevoortia tyrannus*). They have been caught 80 miles from shore off Emerald Bank, Nova Scotia (Vladykov 1936). In the ocean, schooling behavior is the key to the survival of the species; although individuals may be lost to predators, there will always be members of the school that will survive. Because river herring do not have a lateral line, vision appears to be an important aspect of coordinated school movement. Within the schools, the silvery bodies, which are very thin as a result of being laterally compressed, reflect and scatter light. Because they have eyes on the sides of their heads, the light reflection allows individuals to see their neighbors and remain in a tight school. The massive size of the school and resulting light reflection allow the school to appear as one large organism and also makes it difficult for predators to focus on individuals. Thus, the closely knit school moves as one, remaining in coordinated, fluid formation as it navigates through the ocean.

River herring are predominantly planktivores; they are filter and particulate feeders with a diet that consists chiefly of zooplankton (microscopic animals) which they obtain by passing water over gill rakers, tiny specialized combs attached to their gill arches, which strain the water and collect tiny organisms. These organisms are subsequently ingested. Their dietary selections include shrimp-like crustaceans such as mysids, euphausiids (krill), calenoid copepods, and hyperiid amphopods, as well as chaetognaths (marine worms), pteropods (sea snails and slugs), decapod larvae (crab, shrimp, and lobster), and salps (cylindrical tunicates). The efficiency of feeding varies with the size, shape, and density of the prey organisms, the size of the river herring, and the spacing between the gill rakers (MacNeill and Brandt 1990; Palkovacs and Post 2008). River herring can also selectively gulp larger prey (Fay, Neves, and Pardue 1983; Bozeman and Van Den Avyle 1989), using their vision to find shrimp, larger copepods, squid, fish eggs, and even small fish, but this dietary method may be particularly challenging in turbid water.

The schools typically display vertical migration during the day, moving to surface waters during nighttime hours, and plunging to the depths during daylight. This movement up and down the water column is a negative response to light intensity and water temperature, which allows the fish to follow the movement of zooplankton. This type of vertical movement in the water column may also allow fish to avoid predators during the daylight hours and to conserve energy by moving to cooler waters (McKeown 1984).

In the Northeast, it takes about three to four years at sea before the fish become sexually mature. Historical data suggest that alewives can live up to ten years while bluebacks have a slightly shorter life span, seven to eight years maximum. For both species, the female reaches maturity about a year later than the males and females will reach average sizes that are slightly larger than their male counterparts. Non-migratory, landlocked alewives are much smaller than their anadromous counterparts.

From an ecological perspective, there has been a disturbing trend in the stock structure of river herring since the turn of the twenty-first century. The ASMFC river herring stock status report of 2008 indicates that mature fish, both male and female, observed as they migrate to spawning grounds, are considerably smaller and younger than in years past. In addition, many fish that are found on their way to spawning grounds are there for the first time, unlike in the past when most of the fish were repeat spawners. These observations suggest a significant truncation of the age, as well as size structure of the populations and point to some type of stress that may be affecting older, larger fish. In addition to the statistical information from the larger rivers, similar trends have been observed in smaller coastal streams. In a study at Bride Brook, Ct., the run sizes and structure from 2005 to 2008 were compared to data collected from 1966 and 1967. Not only were the 2005–2008 run sizes 20–50 percent lower; the later runs contained smaller, younger fish, primarily first time spawners (Davis and Schultz 2009). Such findings suggest a more fragile population with potentially decreased stability.

Transition to Fresh Water

Movement of up to 1200 miles is an important component of the annual cycle of alewives and blueback herring although very little is known about the cues or conditions that trigger this impressive odyssey. In a seasonal, periodic, and predictable manner, each spring, river herring begin to disperse from their large pelagic schools in a type of movement known as a spawning migration. The river herring are not migrating in a vacuum. Because they serve as forage for other species, the migration and breeding of other fishes is linked to that of the river herring. As summarized by Ames (2004), an 1883 fisheries report indicated that Atlantic cod co-migrated with river herring; thus, early declines in river herring may have contributed to the collapse of the coastal cod fishery in the 1880s.

As they approach the coast from south to north, smaller cadres of the schools peel off into the mouths of different rivers and streams. Alewives were usually seen in the Potomac River by March 4, but did not appear in Canada until April and did not ascend the St. John's River, New Brunswick, until May 10 (Belding 1920). This movement appears to be triggered by the warming of water flowing from the rivers and streams and coincides with the time in which the freshwater rivers and streams are warmer than the temperature of the saltwater habitat of the fish.

What leads marine fishes such as river herring to seek fresh water for their drive to reproduce? This is an important question and it is not unique to river herring. Several theories have been suggested to explain the benefit for fish species of utilizing different habitats during different parts of their life history but the most reasonable explanation stems from considering these fishes from the evolutionary perspective. While many fishes are exclusively marine or exclusively freshwater, there is a surprising number of fish species that have significance tolerance to extremes of salinity and are referred to as "euryhaline" species, i.e., they can thrive over a wide range of salinity. In 1949, Myers suggested the term "diadromy" (McDowall 1997) to label the distinct migration of species between fresh and marine environments, with the underlying assumption that this type of migration is obligate and occurs at specific and predictable stages of the life history of species that exhibit this behavior. Anadromy is one form of diadromy, and describes the life history of river herring in which most of the feeding and growth occur in the ocean while spawning is restricted to fresh water. Catadromy, more typical of fishes in tropical climates, is exemplified by the American eel (*Anguilla rostrata*). Eels spend their adult life in fresh water but migrate to the ocean when they are 5 to 20 years old. The adults eventually, and somewhat miraculously, reach the Sargasso Sea, where they reproduce. The young progeny, glass eels, small and transparent, migrate north along the Atlantic Coast, darken in color and develop into miniature eels known as elvers, which find their way into rivers and streams. The elvers then move up into the freshwater sections of the waterways where they grow and mature, until they are ready to make the return trip to the Sargasso Sea to reproduce and then die.

Of the 250 fish species that have been identified as diadromous (McDowall 1997) it is not always clear whether that particular species arose from a freshwater or saltwater ancestor. Whatever the evolutionary pathway, it is believed that it is beneficial to the species to be able to utilize both types of habitat, thus capitalizing on the prime resources for spawning, growing, and feeding. River herring are believed to stem from a marine forebear, in contrast to salmon, which are thought to have evolved from a freshwater precursor. In the northern latitudes, anadromy predominates because food is more abundant in the marine environment most of the year, but plentiful in fresh water during late spring and summer. Thus adults can utilize the marine food resources, but the young can have access to the abundance of freshwater plankton. Predation may also be a factor that drives the anadromous life history trait. The number of predators, especially those that prey on the young of the species, is potentially much greater in the open ocean than in freshwater environments.

River herring not only make the journey from salt water to fresh water to reproduce; they also make the return journey back to the marine environment after they have completed their reproductive cycle, and many have made this round trip journey five to six times during their adulthood. Such fish species, which have the ability to reproduce more than once, are referred to as iteroparous. This two-way travel is not seen in all anadromous species; although Atlantic salmon and trout in the genus *Salmo* migrate to freshwater spawning areas more than once in their adult lives, Pacific salmon (genus *Oncorhynchus*) are semelparous, i.e., they have a one-way ticket to their freshwater spawning grounds, dying after spawning and thus never making the return trip to the ocean.

There are necessary costs to pay for anadromy, the most significant of which is the stress of moving through waters with differing salinity. The stress is not always apparent; Richkus (1974) noted that juvenile alewives, as small as 25 millimeters (about 1 inch) seemed to be tolerant of salinity changes from 0 to 32 ppt. Although river herring appear to move between ocean and river effortlessly, essential physiological adjustments occur. It has long been noted that river herring do not immediately charge into their freshwater spawning grounds, but, rather, spend some time schooling in estuaries before their upstream journey. Some observers have speculated that this interlude in brackish water, somewhat of a staging period, may be a necessary period of adjustment accompanied by physiological preparation for the transition in salinity. It may also be the case that the time spent in brackish water is a period when the fish are waiting for certain cues, such as water flow or temperature, to migrate upstream. Whatever the reason, when river herring amass in estuaries before their upstream migration, their predators are certain to cue in.

The blood plasma of teleost fishes, i.e., fishes with bones, whether freshwater or marine, is maintained at about one-third the osmotic concentration of the sea (McCormick 2001). In order to preserve this type of osmotic state, fresh water species must take up salt into their systems and retain it while removing excess water. In contrast, marine species drink salt water and must retain water while removing excess salt. The physiological processes responsible for these adjustments are collectively referred to as osmoregulation, and are not completely understood. The gills are the primary site of transport of sodium and chloride, the two ions that comprise the bulk of the "salt" in salt water. Ion transport mechanisms also function in the gut and urinary bladder. Many of the studies on fish osmoregulation have been conducted in the laboratory, using salmon as the model system and it is clear that, in salmon, juveniles develop this capacity for salt secretion prior to entering seawater and that this process is under hormonal control. Scientists perform these studies using a variety of controlled experiments such as administering various agents to fish and testing their salinity tolerance or measuring plasma or blood levels of hormones and other compounds as fish adjust to differing salinities. Generally, laboratory studies point to the conclusion that the process of osmoregulation, as elucidated in salmon, is a phenomenon that is widespread in teleosts.

The Endocrine Influence

Two pituitary hormones, growth hormone and prolactin, play an important role in the process of osmoregulation. In mammals, growth hormone, as its name implies, promotes growth, by indirectly stimulating growth of bones and for its anabolic, muscle-building role, while prolactin stimulates milk production in mammary glands. Curiously, these very same hormones are key components in osmoregulation in fishes. The role of growth hormone in osmoregulation was suggested after it was observed that transgenic salmon, engineered to express large amounts of growth hormone and thus grow to a large size at an accelerated rate, have greater salinity tolerance than wild salmon. It was determined that the salinity tolerance could be linked to the growth hormone itself and was not linked to the larger size of the fish. Some of the actions of growth hormone appear to be mediated by an increase in another hormone, insulin-like growth factor 1 (IGF-1). Both hormones act upon chloride cells in the gill where ion transport occurs and are crucial in the ability of anadromous fishes to acclimate to salt water. Prolactin appears to act upon the gill in an opposing manner by promoting acclimation from salt water to fresh water. Adding further complexity to the hormonal control of osmoregulation, it has been demonstrated that cortisol has a marked effect not only on ion uptake, necessary for life in fresh water, but also on salt secretion, necessary for life in salt water. Thus cortisol may act in concert with growth hormone as well as prolactin in the endocrine control of osmoregulation (McCormick 2001; Mancera and McCormick 2007). Although the details concerning the adjustment are not completely understood, it is generally believed that gearing up for the transition from salt water to fresh water, or in the opposite direction, is one of the physiological costs associated with anadromy. This transition process highlights the potential importance of brackish water systems such as estuaries as staging areas for these biological adjustments.

The Inland Journey

Salmon have been used as a model organism for determining why and how anadromous fishes, migrating to fresh water to spawn, select particular rivers or streams for their upstream migration. Tagging studies, which trace salmon from juveniles to reproductive adults, indicate that they return to the exact location where they were spawned. This mysterious phenomenon is generally thought to be the mode of operation for river herring as well, and has been dubbed the "parent stream" hypothesis, while the process itself is referred to as "natal homing" or "natal philopatry." The ability to return to the parent stream has an obvious benefit. When a spawning area is productive and is a location from which river herring were able to emigrate successfully to the ocean, it is advantageous for individuals to continue to utilize that spawning area. The cohorts may then become adapted to the local conditions in their home habitat and as they and their progeny continue to return to the same spawning area they become a "stock," a subset of the species that is related by virtue of utilization of the same breeding habitat. There is a difference between the concepts of homing and natal homing that is sometimes not distinguished in the literature. Homing refers to a behavior in which mature fish will return to the spawning grounds that they used in previous years, while natal homing refers specifically to a return to the spawning grounds from which a fish originated.

There is ample evidence to suggest that river herring display homing behavior, but, as yet, scientific evidence to support their natal homing has come from indirect observations. Some indicators that river herring return to the same streams to spawn originate in past attempts to restore herring to streams from which they have disappeared. When river herring are taken from one stream and physically transported to another stream system for spawning, the fish will spawn in the foreign stream. Years later, the river herring that returned to these new streams to spawn in spring were presumed to be the three–to-five year old progeny of the relocated stocks (Belding 1920), but they could also be other herring that strayed from a different natal location.

It has been hypothesized that totally accurate homing might present challenges for the survival of stocks if there are major alterations of the natal habitat which may prevent the cohort from reaching spawning grounds and reproducing (McDowall 2001). Therefore, from a survival perspective, it may be beneficial for the species if some individuals are more adventurous and fail to home. Some evidence for straying is provided by observations at restored runs, where fishways have been constructed, impediments to upstream migration have been removed, or spawning areas have been improved. In many of these cases, some river herring begin to use the restored or improved rivers, streams, and spawning habitat within a short period of time. These new immigrants were not spawned in these systems but adapted to take advantage of non-natal habitat. Some researchers have used meristic characters to study homing. Such criteria include different types of counts and/or measurements such as the number of finrays, gill rakers, and vertebrae. The underlying assumption is that fish with similar meristic characters are similar from the genetic perspective and thus constitute a specific breeding stock. After analyzing eight meristic characters of alewives from various areas in the St. John River, New Brunswick, Messieh (1977) concluded that there is considerable straying of fish, particularly between geographically proximate areas, during upstream migration.

Homing has the potential to limit gene flow among members of the species, with local stocks becoming more isolated from the overall population.

The result of such population structuring is the accumulation of genetic differences that offset the stock from the general population. Biologists generally believe that inbreeding and loss of genetic diversity is not beneficial for a species and could be disastrous under certain environmental scenarios. Consequently, from a genetic perspective, it may also be beneficial for some mature river herring to stray in order to mitigate against inbreeding and contribute to the genetic diversity of the species.

Very few studies have examined the genetic evidence for, and possible repercussions of, homing in river herring populations. Using a technique in which different genetic forms of enzyme proteins, known as allozymes, can be detected, Ihssen, Martin, and Rogers (1992) compared alewife populations from different river systems in Maritime Canada. The results of the analysis pointed to genetically distinct populations from the St. John, Gaspereau, and Miramichi Rivers.

During spring migration, Palkovacs et al. (2008) sampled alewives from seven small rivers and coastal streams in Connecticut emptying into Long Island Sound. The populations were compared to one another, as well as to seven nearby landlocked populations. Almost all of the runs included in the study are downstream of the lakes where the non-migrating, landlocked populations are found. Using two different types of genetic markers, mitochondrial DNA and microsatellite DNA, researchers discovered that the alewives from the seven anadromous populations had a low proportion of genetic variability, suggesting the exchange of genes among these populations. From a homing perspective, this indicates that there is somewhat less homing site fidelity than what might be expected if each mature fish returned to its natal stream, and, consequently, all mature river herring may not be homing to natal streams. These findings agree with those of Messieh (1977), which suggest that some straying occurs and that some stock mixing may result from this behavior.

In order to define stock structure on larger spatial scales, three scientists, Jason Stockwell at the Gulf of Maine Research Institute along with Karen Wilson and Theo Willis from the University of Southern Maine, are using morphological and genetic criteria to define stock structure for river herring in Maine. The Maine scientists are also contributing to a major project in which genetic and geochemical markers are being used to study alewives and blueback herring stocks from Florida to Maine (Eric Palkovacs, personal communication). As part of the stock study, Palkovacs and collaborators have reported preliminary genetic findings, using fifteen microsatellite markers, which suggest that homing behavior in river herring has led to genetic differentiation that identifies five regional stocks of alewives (Canada, northern New England, southern New England and New York, Mid-Atlantic, and Carolina), while bluebacks substructure into four stocks (northern New England, southern New England, Mid-Atlantic, and a southern stock) (Palkovacs and Gephard 2012). It is interesting that the study suggested that alewives from Long Island Sound segregated with the southern New England stock while blueback herring from the same area appear to be related to the Mid-Atlantic stock. It is important to note that the genetic stock structure is delineated on a regional scale, not river by river, and that the proposed stock structure does not shed light on the possible mixing of stocks in the ocean,

This research to assess stock structure is not merely an academic exercise; analysis and characterization of stock structure is an important undertaking that has the potential to aid in management and protection of anadromous river herring wherever they are found. For example, if natal homing is not absolute, and river herring stocks in proximate coastal locations have a propensity to stray from stream to stream and reproductively mix, managers have more leeway in using translocation of adults to repopulate depleted runs. The study of stock structure will also shed light on the movement patterns and associations of river herring from particular runs as they spend most of the year in large schools at sea.

Theoretically, from a fish's perspective, a particular spawning area may be excellent, and it thus seems reasonable to speculate that this particular area is being sought by the mature cohort during reproductive events. But it seems quite a challenge for a river herring to find its particular river or stream along miles and miles of coastline. How do they manage to do it?

CHAPTER 3

Home Run

Scientists have identified a number of potential cues that may direct river herring to the perfect spawning location, but overall, the process is still a mystery. What other evidence has emerged for parent stream-seeking in river herring or for homing in general, and what guiding mechanisms are at play?

Heading Home

A type of indirect approach to the study of homing, conducted by Thunberg (1971), relies on laboratory studies in which a number of alewives, taken from one spawning pond, were tested in tanks that had different combinations of source water such as original pond, ponds with different populations of alewives, ponds with no alewives, and water from Narragansett Bay. The directional swimming behavior of the alewives within the tank was interpreted as an indication of water preference and thus, the conclusion was reached that alewives prefer to swim toward water from their natal stream.

A more direct approach to studying homing cues and behaviors involves tagging fish before their seaward migration. Such a feat is easier said than done, and there is some concern that catching and handling fishes, especially sensitive species such as river herring, in order to insert tags, may change the normal behavior of the fishes or compromise their health. Most of the studies have utilized a variety of techniques in which a fish is tagged when it is first caught and then identified via the tag when it is recaptured. In fish migration and homing research, the types of tags have varied but generally include a variety of traditional wildlife identification techniques, including fin clipping, tattooing, branding, inserting chemical tags such as fluorescent markers, radio transmitters, acoustic tags, and small metal or plastic tags. Although tagging has the potential to reveal migratory patterns and other life history traits, some of the methods involve sacrificing captured fishes to look for tags, making the fish more visible to predators, or compromising the survival of the fish in other ways.

Smith et al. (2009) assessed the effect of gastric tagging (radio transmitters inserted through the mouth and into the gut) on alewives and conclude that if a suitable protocol is used, this procedure does not have an adverse effect on the fish. Most studies using such tags have been conducted with shad, primarily to determine how well shad can navigate around dams. Very little information is available about tagging of alewives to understand cues and mechanisms for natal homing. The only available studies involving tagging experiments and large numbers of alewives were conducted in Canada. In five separate experiments conducted over a ten-year period on the lower St. John River, 3,000 tags were applied. Total tag recovery was on the order of 2.3 to 5.5 percent. Recovered, tagged fish were often found in the same St. John River tributaries from which the fish were originally captured (Jessop 1994). It is important to note that in this study, only sexually mature, adult fish were tagged. Thus, using gastric tagging methodology, we have evidence that mature river herring seek the same spawning areas year after year. While the study supports the notion that alewives will return to the same spawning grounds in subsequent years, it does not address the issue of the natal home for these fish.

If river herring are indeed homing to their parent stream or at least to the same stream, what is the "GPS" that they are using? How much of this homeward migration is an active process rather than a less directed movement to a suitable spawning area. In other fish species, celestial, geomagnetic, and olfactory cues have been shown to direct homeward migration. Laboratory studies have also indicated that, in salmon and trout, there may be a genetic contribution to some components of the upstream migration process, most notably orientation (summarized by McKeown 1984).

How does homing work? Several mechanisms have been identified as important components of the navigation and homing system. One such mechanism has been labeled "piloting" and involves specific movement as a response to "a direct sensory cue emanating from the home site that may cause the fish periodically to change compass direction during its journey . . ." (McKeown 1984, 58). By using such sensory cues, migrating fish may gather environmental input that allows them to construct some form of map. Thus it could be postulated that when fish return to their home streams, they are following a map they learned during their downstream voyage from quiet fresh water to the open ocean.

One of the most significant cues for homing may be mediated by the olfactory system, a type of chemoreception, which detects the chemicals that contribute to odors. Within the nares of the fish lie pits containing receptive sensory cells, which detect chemical signals as water moves through as a result of the active forward swimming of the fish or by changes in pit shape as the fish moves its mouth (Reebs 2001). A number of studies with salmon have indicated that young fish can be imprinted and will exhibit homing after exposure to synthetic chemicals, a phenomenon suggesting that although the potential for imprinting is innate, homing to a specific site is a learned behavior. In a very small laboratory study, involving only ten fish, Thunberg (1971) was able to ablate home water preference of alewives by plugging their nares with cotton. When the cotton was removed, home water preference was restored. The exact source of the odors that are the basis for imprinting is still a mystery. Some biologists have speculated that the particular characteristics of the soil and vegetation in the stream serve as the olfactory signals. If this is indeed the mechanism for stream site selection, the olfactory cue or cues, whatever they may be, must be present every year at the time of imprinting. If river herring are using olfactory cues for natal homing, this would mean that the cues must be present during late summer and early fall when juveniles are emigrating from fresh water to the sea, as well as during the inland migration of adults in spring.

Another theory contends that homing is mediated by pheromones, chemicals exuded by other fish in the natal stream (Solomon 1973). Rather than relying on long-term memory, Nordeng (1977) hypothesized that, based on migration schedules, pheromones released from skin mucus signal the homeward migration of salmonids (char, trout, and Atlantic salmon). However, salmonid migration differs significantly from the case of river herring. Mature salmonid movement upstream to spawn coincides with the downstream movement of smolt, young fish, which spend years in fresh water before they undergo the changes that allow migration to the sea. Therefore, mature salmonids may follow the pheromone trail of downstream migrating smolts. In contrast, the young of river herring usually move to the ocean in late summer and fall, about six months before adult spawning migration, so the temporal movement of the year classes would not be expected to overlap. Consequently, the role of pheromones in river herring homing is unlikely.

It is apparent that some studies are in agreement but others present contradictory findings. Homing is a generally accepted concept for river herring, although some recent genetic data point to some straying with subsequent mixing of stocks. Further genetic testing, tagging studies, and other methods of stock assessment have the potential to shed more light on river herring stock structure and afford managers a scientific basis for conservation initiatives to protect our remaining populations of alewives and blueback herring.

The Curious Case of the Landlocked Alewives

The exception to alewives that have a single-minded drive to travel upstream to spawn are the strictly lacustrine (lake-dwelling) alewives, those non-migrating members of the species that inhabit freshwater lakes throughout their lives. If a pelagic marine existence is the norm for most of the life of a river herring, with parent stream selection the driving force for spawning, how did it come to be that members of the species exist as non-anadromous, landlocked populations?

In an examination of the origin of alewives in the Great and Finger Lakes, using a type of genetic marker known as an allozyme, Ihssen, Martin, and Rogers (1992) tested two possible scenarios: an invasion from the St. Lawrence River or entry from the Hudson and Mohawk Rivers via the Erie Canal. The data from the study point to the similarity between the Great Lakes and Finger Lakes populations and a significant difference between the lake-dwelling populations and populations from maritime regions. Furthermore, genetic similarities suggested that the origin of Great Lakes and Finger Lakes stocks may be traced to the Hudson-Mohawk River system, rather than the St.

Lawrence. No matter the origin, the appearance of alewives in the Great and Finger Lakes has had profound effects on these ecosystems.

Using genetic analyses as well as foraging morphology, i.e., size and spacing of gill rakers, gape, and prey selection, scientists examined landlocked and anadromous alewives in Connecticut, and the resulting comparison provides evidence for independent, parallel evolution of landlocked alewives (Palkovacs et al. 2008). A significant finding from this study is the fact that landlocked alewives from different lakes exhibit considerable genetic differentiation indicating that the lake-dwelling populations have been isolated from one another, as well as from their anadromous counterparts. The genetic data suggest an independent evolutionary origin, 300–5000 years ago, for each landlocked population, from a core anadromous population. In postulating the mechanism that drove this independent evolution, Palkovacs et al. (2008) note that there have not been any major geological events within this evolutionary time frame. However, the minimum evolutionary time frame coincides with the period of human settlement of the area and thus, the population differentiation may be the result of human activity, in particular, the construction of dams, which impeded the migration of the fish.

Established landlocked alewife populations can be found in at least nineteen states. The Vermont Department of Fish and Wildlife is carefully watching a population of alewives that was discovered in 1997 in Lake St. Catherine in Rutland County. Their course of entry into the lake cannot be explained from any known natural migration routes, and it is therefore suspected that they were introduced into Lake St. Catherine by accident, from someone's bait bucket or brought in from out-of-state for unexplained reasons by unknown individuals. Alewives were first detected in Lake Champlain in 2003, again in 2004, and by 2005 they became established, thought to have arrived there from Lake St. Catherine. Piles of dead alewives have washed up along the New York and Vermont shores of Lake Champlain. The mass mortality appears to result from the fact that the lake-living alewives are not tolerant of extreme cold, especially for prolonged periods; if there are die-offs under the ice, the fish will pile up on beaches during the spring thaw. They may also be perishing as they move between areas that have wide temperature differences, for example, when they move closer to shore from cooler, deepwater locations prior to spawning. There is concern from sportfishing-related businesses (guides, charter boat captains, bait and tackle shops) that the new arrivals will interfere with restocking of profitable game fish such as trout and salmon, and confound conservation efforts to manage the lake ecosystem.

Landlocked populations of alewives live their entire life cycle in fresh water but adults reach a smaller average size, 15.2 centimeters (6 inches), compared to their anadromous counterparts of about 25.4 centimeters (10 inches). They also grow more slowly, mature at an earlier age, and the females produce fewer eggs. In their lacustrine environments, they inhabit deep, open water but move to inshore locations during spring and summer to spawn.

Because of their feeding behaviors and interactions with other fish species, they have significantly altered lake ecosystems. There is evidence from several lakes that alewives, in general, affect the size and abundance of zooplankton. However, the impact of anadromous alewives, which are transient residents of lake ecosystems is not as profound as that of landlocked fish, which have a permanent impact on zooplankton communities. In lakes inhabited by anadromous alewives, the zooplankton are large-bodied in the spring and smaller in the summer while landlocked alewives cause zooplankton communities to be dominated by small-bodied species (Post et al. 2008; Palkovacs and Post 2008). This difference in prey size has had a pronounced effect on the predator itself; when adjusted for overall size differences, landlocked alewives, which are destined to feed on smaller prey, have noticeable differences in foraging traits: decreases in gape width and gill raker spacing. Palkovacs and Post (2008) have studied this eco-evolutionary feedback and postulated that the smaller prey size has driven the rapid evolution of foraging traits in landlocked alewives.

In lakes, large alewife populations have the potential to deplete food resources for other planktivores and for young fishes of many species, which rely on zooplankton to support the early stages of their growth. Although zooplankton are their dietary mainstay, large alewives will consume eggs and larvae of other fishes and thus have a major impact on native fishes. Competition for zooplankton and predation on eggs and larvae are thought to be responsible for some of the major declines in native species.

Alewives can also impact lakes in their role as a forage base for larger predatory fishes. Larval and young alewives are prey for adult rainbow smelt and yellow perch while mature alewives are prey for many larger freshwater fishes including bass, various types of trout, and salmon. One would think that this level of predation by larger fishes would keep the alewives in check, but in some cases, it has led to even more problems. Consider the case of two unusual afflictions: Cayuga syndrome, which affects salmon, and Early Mortality Syndrome (EMS), which affects trout. Both diseases are transmitted from female fishes to their progeny as a result of very low levels of thiamin, commonly known as Vitamin B1, in the females' eggs. A lack of this crucial vitamin during egg development causes massive mortality of the larvae. As it turns out, the thiamin levels in eggs is so low because female salmon and trout feed on alewives, which, for unknown reasons, are loaded with an enzyme known as thiaminase. Thiaminase, as its name implies, degrades thiamine so that it cannot be stored in eggs. When experiments were conducted to increase or restore thiamine levels to gravid (egg-containing) female salmon 14–23 days prior to spawning by directly injecting thiamine into the body cavity, the mortality of resulting salmon fry was only 2.1 percent compared to saline-injected controls in which mortality reached 98.6 percent. Thus, in the laboratory, direct thiamin supplementation prevented Cayuga syndrome (Ketola et al. 2000). Alewives are not unique reservoirs of thiaminase. Studies have demonstrated that the enzyme is also found in other prey fishes such as rainbow smelt. It is a mystery why alewives and rainbow smelt have such high levels of thiaminase, and even more perplexing is the fact that these fishes are immune from the effects of their own potentially harmful enzyme. Although thiaminase appears to be responsible for poor larval development and mortality of salmon and trout fry, the high level of thiaminase in alewives apparently has no effect on development of alewife larvae or alewife fry mortality.

Alewives have the potential to become a nuisance in lakes if they die in large numbers and wash up along the shore. The cause of alewife mass mortality in most lake systems, fairly common in spring and summer, seems to be linked to temperature fluctuations but may also be due to lack of food. Aside from producing horrible odors, concerns for public health, and closure of stretches of shoreline to recreation, the massive die-offs of alewives lead to decline in prey for larger fishes. The landlocked alewife populations are very resilient; in the lakes where die-offs occur, alewives have been able to rebound fairly rapidly.

Once alewives become established in lakes, there is no way to eradicate them. If some of these populations were established hundreds of years ago, we may even question whether they should still be considered "invasive." Some of the landlocked populations are permanent fixtures of their ecosystems, and it is unlikely that they will ever disappear. Several methods have been used to manage the populations and prevent their numbers from exploding.

The most effective management method has also proved to be a boon for recreational fishermen. Alewife predators, such as nonnative Pacific salmon, have been introduced into lakes, and restocking programs have brought in hatchery-reared trout. The presence of game fish such as Coho and Chinook salmon is the basis of a very large and commercially important recreational fishing industry in the Great Lakes. It is quite ironic that the lakes are being managed for a balance of alewives and Pacific salmon, two nonnative species. It is also ironic that the abundance of invasive alewives in the lakes is in marked contrast to their declining numbers in their natural marine and riverine habitats.

Although most of the landlocked river herring populations are constituted of alewives, there are examples in some southeastern reservoirs where blueback herring have established non-migrating populations. In some cases, the fish were subjected to anthropogenic entrapment; in others, the fish were introduced. A population of non-migrating blueback herring can be found in J. Strom Thurmond Lake, between Georgia and South Carolina. The lake is managed by upstream and downstream dams that have produced the effect of impounding bluebacks. These fish, typically thought of as a cold water species, are subjected to high temperatures and low dissolved oxygen during certain times of the year. The bluebacks were shown to exhibit specific movements within the reservoir as a response to these stressors (Nestler et al. 2002). The inadvertent stocking of blueback herring in Keowee and Jocassee reservoirs in South Carolina was met with the unexpected finding that they were reproducing in the reservoirs, thus forming a non-migrating stock (Prince and Barwick 1981).

Forsythia and Lilacs

River herring movement upstream coincides with other common signs and indicators of spring. Anyone who is interested in observing herring from inland positions in the Northeast can take note of the blooming of spring shrubs. It is commonly known that the best time to see alewives is when forsythia is in bloom, while the shad move upstream with the blooming of shadbush. Blueback herring move inland a bit later in the spring, and they tend to spawn when lilacs are in bloom. The drive to migrate inland arises from physiological factors related to the urge to reproduce. This essentially implies that endogenous transitions and rhythms, tied to gonadal physiology, dictate that the fish must move upstream to spawn, but non-biological factors, such as water temperature, salinity, tides, currents, photoperiod, precipitation, and others, will determine when and how the migration will occur. Thus, biology sets the stage but external stimuli elicit the observed behaviors (McKeown 1984).

The magnet for the upstream migration is a decent current, with fast-moving water, presenting a challenge to which any respectable, mature river herring will respond. The currents result from the flow of water from higher inland elevations to lower elevations closer to the ocean. Currents will be pronounced where narrow waterways meet broader ones, such as where rivers flow into estuaries and where a stream or tributary flows into a river. Such currents may also be more pronounced after heavy spring rains. Water temperature is another important cue. The run usually begins when a few fish, known as scouts, make their way upstream, an event that may precede the bulk of the run by weeks. Wellfleet Herring Warden Jeff Hughes always puts out the word when the scouts arrive. Usually, he has only indirect evidence of the first migrants: the presence of fish scales along the banks of the Herring River or the presence of fish scales in the scat (feces) of river otter or other river herring predators. Eventually, small groups of about five, ten, or twenty will follow, usually after water temperature flowing from upstream locations reaches or exceeds 9–10°C (close to 50°F). In Rhode Island, migration was not observed when water temperatures exceeded 18°C (64.4°F), perhaps suggesting that higher temperatures impose an added stress to osmoregulation (Richkus 1974). In some studies, males have taken the lead and begin to move inland before the females. This may be due to the probability that male gonadal development may be completed earlier than ovarian development (Richkus 1974). Furthermore, the old may be leading the young; while noting the size distribution of alewives in the Annaquatucket River, Richkus (1974) observed that larger fish appeared earlier in the run. Daylight, or darkness, may also be important for the inland trek. Many investigators note that fish move upstream in larger numbers early in the day and in the late afternoon, with a mid-afternoon lull. But there are exceptions to this; in some locations, the bulk of the run may occur during the midday hours or at night, under cover of darkness. This nighttime movement has been noted in Bride Brook, Connecticut, and may be characteristic of other small coastal runs. In some areas, tides may play a significant role in the initiation of upstream migration.

In runs originating on the east coast of Cape Cod Bay, tides can fluctuate by 3–3.7 meters (10–12 feet) so that a low tide can expose very large stretches of intertidal flats, presenting a barrier to migration. When the tide is high, the flats are covered with shallow water and fish can easily find the mouth of creeks and rivers and begin their inland passage.

At certain points along a run, observers can be treated to the spectacle of river herring, courageously and single-mindedly making their way upstream. Sometimes, they are swimming at full throttle, passing by so quickly that they resemble silver bullets. In other areas, where there is slow-moving water or at resting pools, the schools circle as the fish remain in parallel formation. In some locations, they power up impediments such as rocks and natural ledges, against a faster-moving current. People have been drawn to this annual performance—this rite of spring—and it is not surprising to find that our northeastern herring runs were popular tourist attractions (fig 3.1).

In not too distant history, the passage of scouts and early surge of small groups of river herring would be followed by an explosion of hundreds and even thousands of fish. This timing of inland movement has also been noted by sports fishermen. When I went to observe the herring run at the Cape Cod Canal, I passed a fisherman in waders who was making his way to the parking lot with a large striped bass slung over his shoulder. He was struggling under the weight of the fish; it may have been a thirty-five or forty pounder. The stripers sense when the herring are running, and they appear in large numbers as the fish herd themselves into narrow passages to begin their upstream movement. The start of the herring run at the Cape Cod Canal and similar spots where herring aggregate are prime fishing spots in the spring.

Impediments

The last leg of this challenging journey for river herring is full of obstacles, from fast currents which the fish must swim against to impediments such as rocks, downed trees, large tree limbs, and waterfalls. Although river herring can surmount some of these impediments including small waterfalls, they are not as agile as salmon. Rather than aerial, gymnastic leaps, which can be up to 3 meters (10 feet) for salmon, river herring can manage only small waterfalls. Rather than leaping, they garner their strength by resting in slow-moving areas of the river, then accelerate, reach high speed, and swim

3.1. Tourists at the herring run in Taunton, Mass., ca. 1920 (courtesy Old Colony Historical Society, Taunton)

with all their might against gravity. However, gravity will get the best of them when the height of the falls is above 0.3 meters (1 foot). Given the limitations of river herring acrobatics, it is not surprising that the damming and alteration of rivers and streams by humans for industrial uses such as hydropower, irrigation, flood control, and other purposes has annihilated many herring passages as well as spawning habitats and, in other historic runs, caused massive decreases in the number of adult herring that are able to reach spawning grounds. Even if the passage is not completely blocked, some impediments may delay migration or cause migratory fish to travel longer distances and use more energy to reach spawning habitat. The subsequent enhanced depletion of energy resources is likely to contribute to decreased survival of the migrants and potential destabilization of stocks.

Migration may also be delayed by unexpected events such as heavy rainfall. Migrating river herring moved into several "dead ends" during spring 2010 after heavy rains redirected migration paths. Fish became trapped in waterways that were not suitable for spawning, or they were directed to streams where waters suddenly became very shallow or where large waterfalls stopped them in their tracks. In Bournedale, Massachusetts, the Department of Natural Resources scoops up and transports fish from a section of the river near the old Holway axe factory at the Carter Beale Conservation Area where the fish often reached an impasse. Ordinarily, a flood damper control device prevents the fish from moving into this section of the river, which serves as a water overflow area for the dam, but the device sometimes gets clogged and the fish pass over it, or it has to be removed to prevent the flooding of nearby homes. The same heavy rains of spring 2010 produced floods that opened a gate which barred river herring from moving into a flood control tunnel in Weymouth, Massachusetts. Fish were thus able to pass through the gate, and ten thousand or more died upstream, in shallow water, due to lack of oxygen. Luckily, a portion of these wayward fish were rescued by netting and relocation, similar to the method used at Bournedale; this rescue action redirected them to a safer route to their spawning habitat in Whitman's Pond. And those same spring floods washed out a culvert on the Assonet River, which prevented the passage of herring to their spawning grounds.

In some cases, passage may not be impeded but, nonetheless, anthropogenic alteration of the runs can modify the behavior of migrating fish. Many coastal streams flow through culverts under roadways. The culverts vary in size, but they have one feature in common; they create dark tunnels. Herring wardens and observers note that the fish are rather reluctant to pass through these tunnels. I have witnessed alewives lingering before a culvert, circling for long periods of time and making tentative efforts to pass through. Sometimes they retreat, as if looking for an alternative route to their destination or a more opportune time to try again. They seem wary and they often wait for one brave school-mate to be the first one to venture through the dark, covered passageway before the minions follow. These culverts are favorite hiding spots for snapping turtles, predators which can easily make a meal of a passing river herring. In East Weymouth, I saw a large male snapping turtle set up shop under a culvert, no doubt to intercept alewives that pass through the fish ladder in Jackson Square Park. The Herring River in Wellfleet is also home to a snapping turtle that stalks a narrow culvert about one quarter mile before the fish arrive at their final destination in Herring Pond.

Most observers would agree that the runs are not the same as they were in the past. Many aspects of the inland journey of river herring have been influenced by human activity. From the perspective of fish survival, the runs are smaller and the fish are smaller, but from a human perspective, there is not the same keen interest in the spring migration as there was in previous generations. I have met many people who are not even aware of a fish run in their own neighborhood or in a river or stream that they pass over on a daily basis. I don't think the apathy is a result of a lack of concern, but rather a lack of knowledge about the runs due to their overall decline. Irrespective of the attention, or lack of attention, given to their journey, some river herring are still adhering to their annual migration ritual, and, if all goes well, at the end of their long and arduous journey, the river herring will arrive at their spawning grounds and commence their reproductive activities.

CHAPTER 4
River Dance

River herring accomplish an awe-inspiring feat during their spring migration. In some systems, such as small coastal streams, the fish travel only a mile or two upriver, as well as uphill, always against a current; in other rivers, such as the Hudson and Mohawk in New York and the Kennebec and Penobscot Rivers in Maine, the fish manage to travel a good distance across the state, a phenomenal number of miles upstream. No matter the distance, the goal is similar: arrival at prime spawning grounds. The timing will vary along the south-to-north gradient such that blueback herring may begin spawning in Florida as early as January, but it won't be until early May that spawning commences in the Bay of Fundy. Although there may be some temporal overlap in the spawning behavior of the two river herring species, the peak of alewife spawning is usually 3–4 weeks earlier that that of bluebacks. The timing of alewife migration upstream often coincides with the downstream migration of young salmon, called smolts, which have spent the winter in the spawning grounds and have prepared for their migration out to the sea. Consequently, the alewives may provide "prey cover" to out-migrating salmon smolts, protecting the smolts from depredation.

They Come Bearing Gifts

The annual river herring run provides a large surge of marine life into river and stream habitats, with a subsequent impact on the local ecosystem. Not only will river herring reproduce in their temporary freshwater homes, they will also leave other signatures of their visit. Some will die, most will produce massive numbers of eggs or sperm, and all will excrete. This influx of organic material and nitrogen-containing waste products will contribute nutrients to the freshwater habitat. The magnitude of this contribution was studied by Walters, Barnes, and Post (2009). According to this study, 66 percent of the nitrogen in the Bride Brook herring run in Connecticut originated in excretion products while the remainder came from carcasses. The investigators concluded that the influx of nitrogen would not have a significant impact on water chemistry or contribute to eutrophication; it disappeared quickly and was assimilated into the local food web. The same laboratory studied the input of phosphorus into Bride Lake as a result of the annual herring run. It was estimated that the river herring contribute 23 percent of the annual phosphorus load. Although this is a large quantity of phosphorus, it pales in comparison to historic runs in the same system, which were estimated to contribute 2.5 times more phosphorus (West et al. 2010).

End of the Road

Alewife and blueback herring may have overlapping migration routes to freshwater spawning habitats, but each of the species prefers spawning areas with specific characteristics. Two major criteria for spawning site selection are water velocity and substrate characteristics. In general, alewives prefer quiet water for spawning; they often choose inland lakes and ponds as their destination, although they can also spawn in sluggish areas of rivers and streams, generally at depths of 15 centimeters–3 meters (5.9 inches–9.8 feet) (Pardue 1983). The type of substrate in these areas may be sand or gravel, and the bottom may be covered with vegetation and/or detritus. In contrast, blueback herring spawn in areas with higher velocity and, sometimes, deeper water. Because they do not usually ascend to ponds and lakes, they will use tributaries of tidal rivers, channels, and streams. Bluebacks generally select a firmer bottom and are not averse to gravel. They have also been reported to spawn in flooded, low-lying areas in proximity to main streams, and even flooded rice fields (as summarized by Pardue 1983). Because alewives are drawn to ponds and lakes for spawning, they are usually more abundant than blueback herring in systems that have headwater lakes and ponds. As river herring migrate closer to headwaters to spawn, there may be shallower water and more vegetation cover; thus the spawner will be less likely to succumb to predators such as striped bass and predatory birds.

Down to Business

Although many observers have undoubtedly witnessed the spawning of river herring, there are very few written reports of spawning behavior. It is generally the case that older, larger fish are the first to spawn, followed by smaller, younger fish. The larger females, capable of producing more eggs than smaller females, are more likely to be repeat spawners. The age at first spawning and the percentage of repeat spawners generally increases in a south-to-north direction (Pardue 1983). The term "broadcast" is often used to describe the manner in which females shed eggs and males shed sperm. Rather than taking turns, males and females simultaneously release massive numbers of gametes over the substrate. The females may produce tens to hundreds of thousands of eggs in each spawning event. As the females release their eggs, the males will fan their tails to cover the eggs with their *milt* (Belding 1920). The fecundity of river herring, i.e., the production and release of such large numbers of eggs and sperm, ensures the potential to produce a significant number of progeny.

Belding noted that in coastal streams in Massachusetts, many of the females were not ripe during their journey to the spawning area. Their eggs continued to develop to a point at which they were ready to be shed after they arrived at their destination. Belding described the mating system: a group containing one ripe female accompanied by several males (Belding 1920). One female and up to twenty-five males simultaneously broadcast gametes in shallow water or over substrate. This type of spawning activity occurs for two to three days for each "wave" of spawners. In 1977, Loesch and Lund described the spatially separate spawning of alewives and blueback herring in the Connecticut River. Spawning was first observed when the water temperature reached 14°C (57.2°F). The ratio of males to females in the spawning areas was 2:1, and the spawners formed distinct groups containing one female and several males, swimming in circular patterns. It was noted that occasionally a male would nudge the female in the vent with his snout. The group would swim progressively faster, and would eventually dive down and release gametes over the substrate. In shallow sections of streams, the spawners faced into the current and "their bodies distinctly quivered as eggs and sperm were released" (Loesch and Lund 1977). The amount of time that individual fish spend in spawning grounds varies tremendously. There are various reports of river herring spending from only a few days to over eighty days in spawning locations.

The timing of spawning between the two species of river herring is different and appears to be tied to water temperature at the spawning sites. Alewives prefer cooler water and appear to tolerate the cold better than blueback herring. Consequently, alewives spawn earlier in the spring, when water temperatures are a minimum of about 10°C (50°F) in northern latitudes. At the same respective latitude, bluebacks wait until the water warms by at least 5°C (about 9°F). This span between preferred water temperatures translates into a three-to-four week separation of spawning events between the two river herring species. Spawning abruptly ceases when water reaches 27°C (80.6°F).

Due to temperature, and depending on latitude, the timing of reproduction varies. In the St. Johns River in Florida, blueback herring have been observed spawning as early as January. Alewives spawn in the Neuse River in North Carolina from mid-March to late May (Bozeman and Van Den Avyle 1989); alewife spawning commences in Chesapeake Bay from late February and goes into April, and bluebacks spawn from late March until mid-May. In contrast, Maine's alewives will spawn from very late April or early May to early June and Maine's bluebacks from late May to mid-June.

In a study of spawning activity in Occupacia Creek, a tributary of the Rappahannock River in Virginia, O'Connell and Angermeier (1997) noted that both species utilized the exact same spawning habitat, but temporal separation of alewife and blueback spawning followed trends in other areas. During the duration of the study, alewives used the upstream spawning habitat from February 24 to April 15, while bluebacks spawned in the same location from April 13 to May 13. Blueback herring produced more eggs and larvae, but alewives utilized the stream over a longer period of time. O'Connell and Angermeier speculated that this sharing of spawning habitat, near the headwaters of the creek, could be due to both species migrating as far upstream as possible to avoid predation, to avoid competition for spawning habitat with other fish species, or because the site characteristics result in greater spawning success. They also commented that the two species may have been prevented from ascending farther upstream in Occupacia Creek due to impediments such as beaver dams and an old mill dam.

Because of the differences at the ages of maturity, the younger fish that arrive on the spawning grounds are primarily males while females dominate the older age groups. It is difficult to estimate the percentage of fish that are migrating to the spawning area for the first time compared to those which have made the journey in previous years.

Running on Empty

In most cases, river herring accomplish their upstream migration without feeding; they rely primarily on their lipid reserves to get them to their final destination. The energy requirements to produce gametes, i.e., eggs and sperm, while also swimming upstream can impart significant physiological stress to individual fish. Imagine the state of the fish when they have traversed expansive rivers such as the Hudson. Even more astounding is the trek from the Hudson and farther up the largest Hudson tributary, the Mohawk River, a trek that did not occur before the construction of the Erie Canal in 1825. These fish must expend considerably more energy than those that migrate a mile or two upstream. Not only do they have to travel much longer distances, they have to battle stronger currents as they move farther upriver. It is not surprising to find that these extreme distance migrants must replenish reserves along their journey. Examination of gut contents of fish in this system has revealed that some blueback herring in the tidal portion of the Hudson River, and many in the non-tidal Mohawk River, near Rome, New York, had food items in their digestive system (Simonin, Limberg, and Machut 2007).

With the exception of those that travel up the longest rivers, river herring do not refuel during their inland journey and freshwater reproductive period, so that, by the time they reach spawning grounds and shed their gametes, they are literally and physiologically "spent"; they have depleted their energy resources and are ready to head back downstream and pump back up. Observation of the fish on their downstream journey can give a hint at the cumulative biological toll of their reproductive strategy. It is very common to see fish that look "beat up." In late May, I observed spent alewives as they schooled in slow leisurely loops in a resting pool in the Bournedale run. As I watched them circling over and over in the same pattern, I saw many fish with open wounds, as well as some with lesions that appeared to be fungal or bacterial infections. Similar lesions have been observed by others (Hay 1959), and these infections are difficult to overcome in the freshwater environments that the fish had just visited.

Round Trip Ticket

Fortunately for the fish without remaining energy reserves, the downstream journey is usually less traumatizing; the water is usually warmer, and less energy is required for the return trip to the ocean than for the upstream spawning run. But sometimes there are impediments to downstream passage that may delay their return to the ocean and impose further physiological stress on the fish. For example, many of the older fishways were designed for upstream passage, without thought to what the fish would do on the way back. But if the way is clear, rather than fighting currents, powering over waterfalls and negotiating all manner of obstruction, the spent silver bullets can simply go with the flow, leaving the next generation behind to develop on their own and fend for themselves. On their way downstream, they often pass late-run migrants who are only at the beginning of their journey. The adults, having completed their mission, can then focus on feeding in the estuaries on their way back to sea. At this stage, they are ravenous and, in contrast with upstream migration when the only way they can be caught is by trapping or netting, fishermen have caught spent river herring in estuaries on hooks with artificial flies. As with the cues involved and the manner in which predators intercept them during the upstream migration, predators will gravitate to areas traveled by the spent adults on their journey back to the marine environment.

Local Conditions

The ecological, physical, and chemical features of the spawning area are not only important for spawning; these features are critical to the fertilization of gametes and the subsequent development of the next generation of river herring. There are several parameters that function in this regard: aside from water velocity, cover, and substrate optimal for spawning, described above, water flow, temperature, salinity, amount of oxygen, and pH also have an effect on the success of reproduction by impacting eggs and larvae. Many of these parameters, measured in several studies, have been summarized by ASMFC (2009). Eggs and larvae must be well oxygenated in order to survive. Lack of oxygen (anoxia) is a sign of poor water quality, often a result of human activities such as agricultural run-off or sewage discharge. The minimal oxygen requirement for the survival of alewife eggs and larvae is about 5.0 milligrams/liter. Most eggs and larvae are found in fresh water where the amount of salt is very low, less than 0.5 ppt, and have optimal survival at salinity of 0–2 ppt. A number of studies suggest that an optimum pH range for survival of eggs and larvae is 5–8.5. In laboratory studies of alewives from a Virginia stream, pH values between 5.7 and 6.5 were reported to be lethal (O'Connell and Angermeier 1997), while in small coastal streams in Massachusetts, pH 8.2 was shown to support optimal larval growth (Kosa and Mather 2001). River flow can also have an impact on the survival of river herring eggs and larvae with the two extremes of flow, drought and very high velocity, posing different hurdles. Drought conditions have the potential to alter water chemistry and temperature, while very high flow may disperse eggs to suboptimal locations and also create turbidity, which may make it difficult for the larvae to find food.

Human populations and activities have made their mark on river herring reproduction in many ways. The most obvious anthropogenic effect is obstruction to spawning grounds, but other impacts are subtle. Limburg and Schmidt (1990) assessed spawning activity (predominantly alewives) in sixteen Hudson River tributaries. The mouth of the Hudson is in New York City, one of the most densely populated cities in the country. As one moves upstream along the Hudson, there is less and less human habitation and development in the watershed; thus an urban gradient is formed along the river. It was hypothesized that there would be a correlation between the urban gradient and the extent of spawning activity. The investigators found that an upstream directed "spawning front," i.e., spawning activity, as measured by numbers of eggs and larvae, was not graduated, but rather, localized to specific upriver areas, farthest from urban centers. Thus, it seems there is a threshold effect in which a certain degree of urbanization precludes alewife reproduction in Hudson River tributaries. Although salinity was ruled out as a factor to explain the localization of the eggs and larvae, other factors associated with urbanization, alone or in combination, may play a role such as habitat alteration, siltation, altered levels of dissolved oxygen, and pollution. These anthropogenic effects are not limited to urban development; agricultural use may also pose problems for river herring reproduction due to runoff into the watershed, an event, which is more likely to occur during heavy rains or spring snowmelt.

Eggs and Larvae

The eggs of river herring are 1.3 millimeters or 1/20th of an inch in diameter, 0.8–1.27 millimeters for alewives, 0.87–1.11 millimeters for bluebacks (Bozeman and Van Den Avyle 1989), barely visible to the naked eye, but there are many of them. The unfertilized eggs of the alewife are green in color, but turn amber when fertilized, while those of blueback herring are amber when unfertilized and then turn more yellow after fertilization. The eggs are filled with small oil droplets (Bozeman and Van Den Avyle 1989). There are various reports in which the numbers of eggs shed by female river herring have been counted, and the numbers are quite large. For example, in alewives, a range of 60,000 to 100,000 eggs per female was reported for Chesapeake Bay, 48,000 to 360,000 eggs per female for Bride Lake, Connecticut. Blueback herring females were reported to shed 45,000–350,000 eggs in the Connecticut River system (Pardue 1983). Although, in general, larger females are capable of producing more eggs, one study conducted in Georgia provides evidence that as females age, their fecundity decreases. In a Connecticut study the number of eggs in ovaries were tallied as a function of fish size/age. It was reported that fecundity generally related to the length of the fish but declined in the larger females (Loesch and Lund 1977).

The convergence of egg and sperm with subsequent fertilization follows the mass spawning events orchestrated by river herring at their final destination. If the eggs are deposited in still water, they will tend to be demersal for several hours, i.e., float to the bottom where, initially, they may be quite adhesive, adhering to one another as well as sticking to the substrate. If the eggs are shed in moving water, they will remain in the water column and not have the opportunity to sink. The eggs undergo a process known as water hardening; they lose their adhesiveness and become pelagic. As the fertilized eggs float through the water, development of tiny fish begins with a stage called the yolk sac larvae, 2.5–5 millimeters (0.1–0.2 inches), which use the nutrients in the yolk to grow and develop.

The larval stage of river herring is generally two to three weeks. The resulting larvae will develop in the fertilized egg until it is time to hatch, an event which is very temperature-dependent: the warmer the water, the quicker the young fish will hatch. Laboratory studies of Hudson River alewives, in which the adults were acclimated to 18.6°C (65.5°F), indicate that the process may take fifteen days at 7.2°C (45°F) while temperatures of 28.9°C (84°F) will lead to speedy development lasting a mere two days. At an intermediate temperature of 15°C (60.1°F) baby alewives will be ready to hatch in six days (Kellogg 1982). Not only does temperature affect the time that it takes for a fish to hatch, it also has an impact on the hatching success. The water can't be too cold or too hot as 100 percent mortality of larvae occurred when water temperatures exceeded 32–33°C (89.6–91.4°F) for twenty-four hours. In additional experiments with Hudson River alewives, maximum hatching success occurred when water temperature was 20.8°C (69.4°F) and no hatching occurred when water temperatures reached 29.7°C (85.5°F). The best larval growth occurred at 26.4°C (79.5°F) (Kellogg 1982). It is important to note that these average temperatures are likely to be dependent on the temperature at which the adults were acclimated and furthermore, the data were collected from experiments using only a few female alewives as the source of eggs; the range in hatching temperatures from eggs obtained from these females was from 15°C (59°F) to 26.7°C (80°F). It is significant that the optimal growth for larvae exceeds average spawning temperatures by at least 10°C (18°F). Therefore, rapid warming of the water in the spring may actually be beneficial. Although larvae cannot tolerate salt water, several studies suggest that they may grow slightly faster if the water is slightly saline.

Generation Next

At the hatching stage, the tiny fish are about 20 millimeters (0.78 inches). After hatching, the yolk sac is resorbed, a process that may take 2–3 days for blueback herring, 2–5 days for alewives. The tiny fish are "positively phototropic," i.e., they are attracted to light and are thus found near the surface, often in shallow areas, where they begin to eat and grow at a rapid rate. Pardue (1983) speculates that the signals involved in the timing of reproduction are geared to the annual plankton production cycle. If the timing for the production of larvae is not synchronized with production of plankton, the subsequent development of juveniles is in jeopardy, possibly accounting for observed deficiencies of entire year classes within stocks. Tiny translucent river herring are food for many fresh water fishes. The larvae are vacuumed up by white and yellow perch, suckers, and minnows. If the river herring manage to survive the earliest stages of development and become small fry, they may still succumb to predation by bass, perch, trout, and pickerel. If they survive the freshwater predator gauntlet, saltwater enemies will await them when they migrate to estuaries and the ocean. Hence, fecundity is a hedge against the low probability of an individual larva becoming a member of the "young of the year." By the time the developing herring is about 30 millimeters (1.2 inches) it actually resembles a tiny version of the adult, and scales have appeared, although the abdominal serrations have yet to form (Belding 1920). The tiny fish are miniature replicas of the adults when they are 45 millimeters (1.8 inches) long (Pardue 1983). Most are not yet ready to join the adult population. They will spend several months bulking up in nursery areas before they take the oceanic plunge.

CHAPTER 5
Cool Running

As the young river herring continue to feed and grow, they pass their time in areas that are rich in zooplankton. They are gearing up for a major habitat shift, from their relatively confined natal lake, pond, river, or stream to the vast and open ocean. Some investigators have attempted to track their movements during this life history stage and have found that juveniles may not remain in the area where they originated, but instead, vary their behavior and location during this early period.

Nursery School

Some alewives, having begun their development in headwater ponds, will remain in this location throughout their growth period if food is plentiful. Pardue (1983) estimated that the minimal concentration of zooplankton needed to support optimal growth of juveniles is 100 per liter. River herring, whose juvenile stages are exclusively in rivers and streams, have more diversity in their travels with movements upriver, downriver, and even into brackish water. Behaviors also differ if we look at juveniles from the south compared to those in more northerly areas. In North Carolina, juveniles may move to lower reaches of the rivers in which they were spawned, into tidally influenced freshwater or even as far as slightly brackish water habitats. In the Neuse River in North Carolina, juvenile blueback herring were found in areas with very little current, very low salinity, and mud or detritus substrate such as water draining hardwood swamps. Some of these blueback herring moved and overwintered in shallow estuaries with higher salinity, which they utilized as secondary nurseries (Bozeman and Van Den Avyle 1989). In Chesapeake Bay, juvenile alewives use a variety of habitats: they live in freshwater tributaries in spring and early summer but then they may move further upstream in mid-summer as the water becomes more saline or as juveniles move from smaller tributaries and side channels into the main river. Others will remain in brackish water throughout the summer. Perhaps as a mechanism to mitigate against competition for zooplankton, the two species of river herring juveniles do not mix. In the Potomac River, juvenile alewives were found mostly along the surface during the day, but as summer progressed into fall, they moved to deeper water and eventually were found on the bottom. In contrast, blueback herring were usually found nearer the surface throughout their stay. Alewife and blueback herring juveniles were noted to partition in the water column at night with alewives at mid-depths and bluebacks at the surface. Somewhat similar partitioning was observed in the Hudson River, but instead of depth, the fish moved laterally in the water column. Alewives were found near shore at night, while blueback herring moved in-shore in the evening. In addition to spatial segregation, juvenile alewives and blueback herring may also avoid competition by consuming different types and sizes of prey. Food may be partitioned between the two river herring species by virtue of the fact that alewives spawn 3–4 weeks earlier than blueback herring and thus develop into juveniles earlier and achieve larger sizes that the bluebacks in the same river systems. This decreases interspecies competition because the young alewives will consume larger prey items than the blueback herring young when both are feeding in the same waterway. Many of these studies, conducted during the 1970s and 1980s are summarized by Fay, Neves, and Pardue (1983).

In headwater ponds, the tiny fish somehow find one another and begin to practice their schooling behaviors. On crisp October days, observers can witness a fish ballet in the shallow regions of Higgins Pond at the Sluiceway where it connects to Gull Pond in Wellfleet. River herring fry move in synchrony, forming circles, ovals, and other figures, almost like calligraphic letters, which seemed to spell out that they were preparing to leave as a group. The schools are small at first, but their numbers increase over the course of a few days to a few weeks, from dozens to one hundred or more, as others join the formations. These schools represent the nursery school graduates, ready for cues to make their first significant move and begin the next phase of their development.

Waiting for a Sign

Throughout summer and early fall, when they are 5–10 centimeters (2–4 inches) the young of the year begin to descend from the nursery to the ocean. The young leave over a long time span: between June and early November, depending on the latitude. What are the cues that tell them that it is time to move on? Richkus (1974) found that juveniles migrated down the Annaquatucket River drainage system in North Kingston, Rhode Island, as a response to environmental factors such as nursery pond volume outflow, plummeting water temperatures, rain events, and possibly crowding. Similar factors are also important in other runs that have been systematically observed. Precipitation, with a concomitant increase in the volume of the nursery ponds, increases downstream water flow and offers a quick current ride to the ocean for juveniles. However, if the flow is too rapid, such as above 10 centimeters per second (4 inches per second), which may occur in narrow channels, the young fish will hold back. Larger rivers can support the flow of much higher volumes of water, which may pass in short periods of time but without significantly increasing the overall velocity (Fay, Neves, and Pardue 1983). A decrease in water temperature and crowding may lead to lower amounts of zooplankton and, consequently, competition for food. Several investigators suggest that it is actually the decrease in food availability that serves as the emigration signal, causing large numbers of juveniles to depart suddenly (Yako, Mather, and Juanes 2002). A single factor, or more likely some combination of the above factors, may jumpstart individual small schools of juveniles to leave their nursery and begin their reverse migration. Although emigration has been observed in daylight hours in some systems, the ability to avoid predators, afforded by dark nights, may also tie emigration to new and quarter moon phases (Yako, Mather, and Juanes 2002).

Arrested Development

Juveniles do not all follow the same emigration trends. In certain North Carolina sounds, and in the upper reaches of Chesapeake Bay, there have been reports of cohorts of juveniles that overwinter on the nursery grounds or in brackish water near the coast, dispersing to the ocean as early as the spring of the following year, coincident with the spawning migration of adults. Little is known about the survival value of this delayed emigration strategy. Another case of delayed emigration was reported by Walton (1983) for alewives of Walker Pond, Maine. The Walker Pond alewives persisted in the larval stage for extended periods, remaining as larvae until mid-August, when a comparison group in Damariscotta Lake had already completed their transition into juveniles. The Walker Pond alewives eventually developed into juveniles but only attained dwarf status by fall of their first year. The alewives that emigrated from Walker Pond were larger than other Maine stocks, and their scales displayed a growth check, indicating that the alewives were leaving the pond in their second year, at about sixteen months of age. The reasons for this delayed emigration strategy of Walker Pond alewives is unknown; however, it was noted that Walker Pond is in proximity to areas where copper, zinc, and lead were mined. Perhaps the metals had a direct affect on the development and growth of the fish or the concentration of zooplankton in the pond (Walton 1983), and so the fish remained there until their second year, when they achieved a larger size.

Seaward Emigration

The first seaward emigration is a critical juncture in the life history of river herring. Young of the year that leave their natal lakes, ponds, streams, and rivers, and complete their passage to the ocean surmount a considerable hurdle in their lives. Depending on local conditions, many fish within a year class may be lost at this stage. The patterns of emigration may vary when small coastal streams are compared to the larger riverine systems. In small streams, water flow tends to decrease over the course of the summer and early fall, so if the juveniles don't leave in a timely manner, they run the risk of being trapped as water temperatures and food supplies decrease. This scenario is generally not the case for river herring that have been spawned in large river systems.

Many studies have focused on the actual timing of emigration of juveniles. Rather than a continuous outpouring of young river herring, the bulk of the emigration occurs in discrete time periods. In a coastal Rhode Island stream, Richkus (1974) found that 60–80 percent of fish migrated downstream in a series of intermittent pulses, rather than as a single large wave and that the movement occurred on a limited number of days. More recent observations in small coastal systems show the same trends. In a study of emigrating river herring juveniles in the Herring River in Bourne, Massachusetts, which empties into the Cape Cod Canal, alewives displayed a bimodal emigration pattern, with an early migrating group in July and August, as water rose to peak seasonal temperatures, and a late migrating group, in November and December, as waters temperatures were precipitously declining. The emigration cues may be different for each group. The earlier group may leave to avoid potentially lethal high temperatures or competition for food, while the late fall group may leave to avoid potentially lethal low temperatures. It was interesting that the researchers who conducted this study found that the earlier emigration wave consisted of significantly smaller fish, and suggested that early emigration at a smaller size was a trade-off; the smaller, earlier emigrant alewives would not have to compete with the later emigrants or blueback herring for food and could get access to the increase in food resources that the ocean offers. During the study, Bourne's Herring River had a steady flow, so entrapment of some of the alewives was probably not a factor; thus the late emigrants may have used a "size maximizing strategy." By growing larger than the earlier emigrants, they are potentially more predator-proof on their downstream journey and also at the time of their arrival in the ocean. In contrast to young alewives, the same study noted that young blueback herring left the river in a single peak, from late September through October, coincident with declining water temperature, the appearance of a new moon, and perhaps the attainment of a threshold size (Iafrate and Oliveira 2008). Many observers have noted emigration to be coincident with moon phase, but it is not clear if a new moon, which could potentially offer more predator protection due to darker nights, is a true cue or is a consequence of timing due to other cues.

Three emigration pulses were documented in the well-characterized alewife population of Bride Lake, Connecticut, extending over mid-June to mid-August, a period when zooplankton are still plentiful (Post et al. 2008). These pulses of one to two days duration accounted for emigration of 80 percent of the juveniles in a single year class. (Gahagen, Gherard, and Schultz

2010). During the observation period, the emigration pulses were associated with three parameters that may individually or collectively contribute to the emigration signals: precipitation events, increases in stream discharge, and transient decreases in water temperature. In the first four weeks of emigration, the alewives left Bride Lake at dawn, but later in the season, during the last four weeks, the alewives left at night. The scientists speculate that outmigration under the cover of darkness may occur during the late migration period because the water is shallower, flow is decreased, and vulnerability to avian predators may be higher.

There is a suggestion that young river herring may not wander far from shore when they first emigrate to the ocean. In winter bottom trawls from December through April, 1972–1975, immediately off the coast of southern New Jersey, 79 percent of the catch consisted of blueback herring, 16 percent alewives, and the remainder, shad. Analysis of size of the catch indicated that most of the fish were age 0+, i.e., young of the year that had recently left estuaries of rivers or bays. Although the origin of these young fish was not determined, it was speculated that these areas, up to 8 kilometers (almost 5 miles) from shore, were the overwintering grounds for young river herring (Milstein 1981). These waters were at the outer limit of estuarine influences, and thus were slightly warmer (4.5–6.5°C; 40–43.7°F) and had a higher saline content (29–32 ppt) than the nursery areas in the estuaries. In some very large watersheds, such as the Delaware River, Chesapeake Bay, and the Connecticut River, juvenile river herring may overwinter in deep estuarine waters.

Period of Adjustment

Salmon are generally used as a model system for the remarkable physiological process whereby anadromous fishes make their transition between fresh water and salt water. This adjustment process occurs as adults migrate to freshwater spawning areas, and then when juveniles travel to the ocean. In contrast, adult river herring make these adjustments twice each year: once during upstream migration and again when they return to the sea. As juveniles, river herring make this transition for the first time, during emigration to the ocean. Is this adjustment more difficult or more stressful because they are smaller than adults? Is the physiological adjustment process the same for the young of the year as it is for the adults? Stephen McCormick of the Conte

Anadromous Fish Research Center in Turners Falls, Massachusetts, has been at the forefront of this research. Many of the findings from his laboratory, regarding salmon and shad osmoregulation, can be generalized to river herring. The exact timing of the development of salinity tolerance in shad was studied under controlled laboratory conditions and was found to be maximal at 58–127 days post hatch (Zydlewski and McCormick 1997). If this timeframe also applies to river herring, this would translate to a period that corresponds to about 2–4 months of freshwater growth. The study also revealed that survival in seawater for 24 hours is a good indicator of long-term survival of the juveniles. A review of events that can affect osmoregulation in juveniles identified several parameters that can have serious negative impacts on preparation for emigration (McCormick et al. 2009). If osmoregulatory mechanisms are somehow compromised, the juveniles will have a decreased ability to make the necessary adjustments, increased susceptibility to diseases, and a decreased chance of survival. The parameters of interest are mainly those found in headwater spawning areas. There is a water temperature range which is optimal for the changes that occur to facilitate osmoregulation and if the temperature is too high, there may be difficulty with osmoregulation. Another important parameter that influences the ability of juveniles to osmoregulate is pH; when pH decreases, aluminum, an abundant soil component, leaches into the water and may inhibit the physiological processes required for osmoregulation. Even when low pH is not a year-round issue, spawning areas can become acidified as a result of certain events such as spring snowmelts and fall storms. Contaminants from the environment, such as endocrine disruptors, may be particularly harmful to juvenile river herring, which depend on a balance of hormones for osmoregulation, and this balance may also be important for imprinting (McCormick et al. 2009).

The Bone That Tells a Story

Presumably, it is before or during their emigration when juveniles become imprinted and learn to recognize and respond to environmental cues that allow them to return to their birthplace as adults. As river herring develop from larvae to juveniles and through their adult stages, whether they are in fresh water or the ocean, they pick up traces of their environment through their food sources and the chemical signatures of the waters they inhabit. Many in-

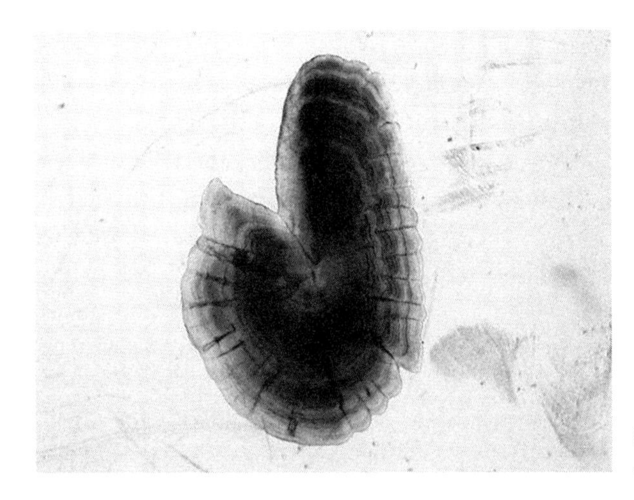

5.1. Otolith (courtesy Jason Stockwell)

vestigators are using a high tech approach to sorting out river herring stocks by employing the analysis of a tiny bony component, called the otolith, in the inner ear of the fish. These calcareous structures form early in the development of fishes and grow continuously throughout their lives. The otolith enlarges, layer upon layer, and as it does, discrete concentric rings are formed (fig 5.1). Some of the rings are dark bands and represent yearly growth rings, similar to the rings in a cross section of a tree trunk. Thus the age of the fish can be discerned. Why use the otolith when it is much easier to remove scales and read the growth rings to infer the age of the fish without having to sacrifice the fish? The otolith can tell more about the fish's life story than scales alone. Not only can prominent annual growth rings be seen; the rings take on a shape that may be characteristic of fish that share the same habitat. In addition, the otoliths display unique microstructure, i.e., narrower growth bands within the annual growth ring, that represent daily growth patterns: patterns that are established as a result of the feeding intervals and the environment, patterns that may reflect the temperature and condition of the waters through which the fish has traveled (Dufour et al. 2008; Elsdon et al. 2008).

In order to read the otolith and unravel the life history of individual fish, the fish must be sacrificed. I visited Jason Stockwell's lab at the Gulf of Maine Research Institute in Portland, Maine, on a day when otoliths were being extracted from river herring, mostly alewives but a few bluebacks in the lot, collected from various sites in Maine and being prepared for examination. Each fish was weighed and measured and digitally photographed so that specialized software could be used to take numerous fine scale measurements. Interns Kim Little and Kaylyn Becker were assisting Research Technician Zack Whitener as the fish were processed and the otoliths removed. The gonads were excised, staged according to the level of development, and weighed. Then Whitener made some expert surgical slices to reveal the minute otoliths, measuring only a few millimeters, on each side of the head. The extracted otoliths were then rinsed, dried, and prepared for microscopic examination. After the delicate extraction was complete, the fish remains were destined to serve another function: bait for lobster fisherman.

I was able to observe a processed otolith through the microscope and, with assistance, could make out the annual growth rings, but the finer structural details and more subtle growth increments appeared very complex. With thousands of fish in the queue to be processed, it looked as though the lab would be busy extracting otoliths all summer and examining them all winter. As Stockwell explained, the overall goal of the study is to interpret the otolith patterns as a "natural tag," because fish that experience the same environments as they grow and migrate share similar otolith patterns. Thus, adult fish within the same cohort that have migrated to the same spawning lake and have similar otolith microstructure within the early annual growth rings most likely shared the same environment during their first year of life and most likely originated in the same spawning grounds. In addition to otolith microstructure, the microchemistry of otoliths can be used to analyze stock structure and natal origins. The ratios of trace elements such as calcium, barium, strontium, and magnesium in segments of the otolith reflect the ratios of those elements in the environment in which the herring spent the portion of its life cycle when that part of the otolith was formed. The analysis of otolith microstructure and microchemistry has the potential to identify and group fish that consistently return to the same spawning lakes and thus define the stock. In addition to the otolith "signatures," Stockwell and collaborators are also using morphometric (size and shape) and genetic data obtained from each fish to assist with the objective of grouping fish into separate stocks.

This complex analysis also has potential to provide direct evidence for natal homing by designing and employing a longitudinal study in which some fish are sampled within their first year, before they leave their spawning grounds. Their otoliths can then be compared to those of adults that return to the same lake to spawn, three, four, or five years later. However, a serious drawback to this approach was described by Gahagan et al. (2012) when they analyzed otoliths from alewives and blueback herring from coastal rivers in Connecticut. The investigators were not able to confidently assign individuals to their natal sites due to the lack of unique differences in water chemistry among the sampled sites and evidence that some fish were very mobile at early life stages.

Although not much is known about the growth and survival of the young of the year from the time they finally enter the ocean to the period in which they return to fresh water to spawn, these tiny ear bones will serve as tracers and their analysis will begin to fill in some of the gaps in our understanding of the movements of the young fish and the structure of local stocks.

CHAPTER 6
Fishy Business

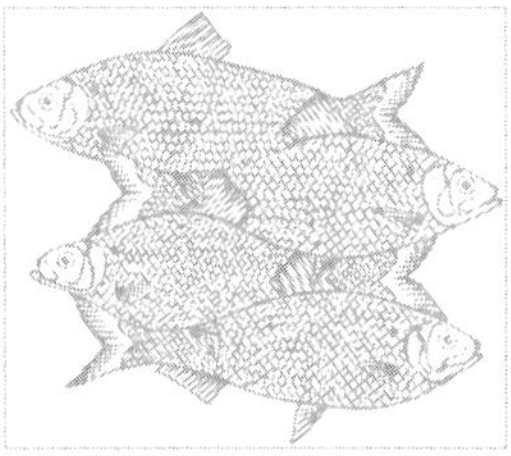

Dating back to colonial times, human inshore activities have greatly impacted river herring stocks. From the early 1600s, after Squanto instructed the settlers how to catch migrating herring and use them for food and fertilizer, until the early twenty-first century, many coastal towns have had some type of herring fishery. In 1623, when Maine was still part of Massachusetts Bay Colony, Article 8 in the Colonial Fishing Laws related to Inland Fisheries of Massachusetts (1623–1886) provided the right to catch fish, along with the right to hunt and the right to fowl, but left the door open to further regulation. Article 8 stated "that fowling, fishing and hunting be free to all the inhabitants of this government: provided, that all orders from time to time made by this General Court for the due regulation of fishing and fowling be observed in place or places wherein special interest and propriety is justly claimed by the court or any particular person."

Initially, anyone could catch an unlimited supply of river herring for their own use, simply by using a net to scoop them up in narrow stretches of streams where they would pass on their way to spawning grounds. With time, towns began to regulate the fishery by leasing fishing rights, or operating it directly as a means to generate revenue. As human settlements became larger and the need arose for energy to power grist mills, clothing mills, foundries, electricity, and other enterprises, rivers were harnessed in a manner that presented obstacles to migrating river herring. Some of the environmental costs of these progressive initiatives were not immediately appreciated.

Where Were the Weirs?

In 2008, after the removal of the Fort Halifax Dam, the remnants of an ancient Native American weir, a type of fishing trap, was uncovered by the retreating waters at China Lake Stream, which flows into the Sebasticook River in Winslow, Maine. Prior to this time, the Fort Halifax Dam impounded the China Lake Stream and caused the water to be at flood heights, thus covering the weir. Weirs such as this were very common and were made from large river stones, which were piled into a wall-like structure that would trap or slow down the upstream migration of alewives. These weirs were widely used by Native Americans throughout the Northeast. At Town Brook, which flows 1.5 miles from Billington Sea to Plymouth Harbor, where settlers of Plymouth first learned to catch alewives from Squanto, the Natives would use a "heap of stones" to form temporary traps to slow down the upstream momentum of river herring. When the river herring rested in the pools formed by the rocks, they could be easily caught. A visit to the colonies in 1674 resulted in the following report written by John Josselyn: " . . . they come into fresh Rivers and Ponds; there hath been taken in two hours by two men without any Weyre at all, saving a few stones to stop the passage of the River, above ten thousand" (Josselyn 1988 [1674], 221).

Alexander Young described a weir fishery that operated near Boston in 1633. He was describing the fishery on the Charles River in Watertown (Water-towne) and nearby Newton (New-Towne): " . . . a great pond, which is divided between those two towns, which divides their bounds northward. A mile and a half from this town is a fall of fresh waters, which convey themselves into the ocean through Charles river. A little below this fall of waters, the inhabitants of Water-towne have built a wear to catch fish, wherein they take great store of shads and alewives. In two tides, they have gotten one hundred thousand of those fishes. This is no small benefit to the plantation" (Young 1846, 403).

Native Americans also erected weirs on other rivers such as the Taunton, which, together with its tributaries, supports the largest herring run in the state of Massachusetts. Each spring, the Wampanoags would travel from near and far, some arriving from areas of Mount Hope Bay, in what is now the town of Bristol, Rhode Island, to set up camps in fishing grounds such as Cohannit, which is currently part of the city of Taunton. Here the river was narrow, and fish could be trapped in weirs that resembled a cage. They were made from closely spaced wooden poles, stuck into the bottom to form a fence-like structure. When the area was colonized by Europeans, the settlers continued the tradition of trapping alewives at Cohannit, and renamed the settlement Weir Village. Taunton became famous for its herring runs and was even dubbed "Herring Town," after the herring runs became a source of revenue (figs. 6.1 and 6.2) and a feature of local tourism. A 100-foot-long section of the run near the former Weir Village shunted the herring into man-made pools that would allow the fish to pass upstream and bypass Nemasket Mills, where part of the river was dammed to supply power to the cotton industry. The mills, which coexisted with the local herring fishery, were of great economic importance to the area, employing three hundred persons to

6.1. Hauling in the river herring seine net, Taunton, Mass. (courtesy Old Colony Historical Society, Taunton)

6.2. River herring harvest, Taunton, Mass. (courtesy Old Colony Historical Society, Taunton)

produce thirty-five thousand spindles of cotton each year. The pools formed a spectacular viewing spot (now part of a housing development known as Riverbend Condominiums) that, each spring, attracted thousands of visitors who might pass a Sunday afternoon with a picnic lunch, watching the herring swim upstream (see fig 3.1). In the local shops, tourists could purchase herring to eat as well as a multitude of river herring souvenirs, including a book shaped like an alewife. The Taunton run inspired several homages to river herring such as a poem by a New Bedford native, Thomas J. Taylor, entitled "Our Finny Visitor, A Song of Appreciation," which includes the refrain:

> The multi-boned herring, unique-
> flavored herring,
> The famed Taunton herring that
> Runs in the spring.
>
> —Poem courtesy of Old Colony Historical Society, Taunton, Mass.

Another tribute was composed in 1905 by F. Bosworth Melvin:

SOUVENIR OF TAUNTON RIVER

When God the Creator created fish
He made some of excellent kind,
Which scattered broadcast around the world,
Here, as Taunton Herrings we find,
Maybe deep in the warm Gulf Stream they play
While the winter days go by,
For they always come from some unknown home
In Springtime, when summer is nigh.
Through the bay that washes the pebbled strand
That borders the green sunny slope
Where Metacomet, King Philip,
Sat on his rocky throne at Mount Hope.
Past the "City of Spindles" of great renown
Where cloth for the nation is made
The herrings speed on where the tide runs strong
Nor rest neath the bridge's shade.
Not the whistle shrill of the "old nail mill,"
Nor the steamboats rough commotion
Will turn them back from their perilous track
When onward is their notion.
Methinks, somewhere near Conspiracy Point
A halt in their run they make
The whetting their fins on The Old Whale Rock,
Decide Taunton main to take
Some late in their run, in frolicsome fun
Near the river bank will hark
To musical sound of Merry-go-round
Stationed at Dighton Rock Park.
No wonder the shouts of fishers resound
From Narrows to billowy bay
When getting good hauls of plump shining fish
At their stations along the way.

—Poem courtesy of Old Colony Historical Society, Taunton, Mass.

One of the major tributaries of the Taunton is the Nemasket River, which winds through the towns of Middleborough (sometimes spelled Middleboro) and Halifax and has its headwaters in the Assowompsett Pond Complex in Lakeville, the largest natural freshwater body in Massachusetts. *Nemasket* or *Namassakeesett* is a Wampanoag word for "place of fish," and there is archeological evidence to suggest pre-contact Native American villages or encampments on the Nemasket that date back thousands of years. The Natives would capitalize on the spring fish frenzy and gather enough river herring to fertilize their spring crops with some left over to preserve for winter sustenance. The Native population was somewhat nomadic, staying at the inland fishing camps such as Muttock (or Muttuck) and Titicut for the spring herring run and then moving to summer coastal camps to take advantage of the offerings of the marine environment. Muttock is now the site of Oliver Mill Park in Middleborough, where herring can still be seen each spring. When the Natives were driven away or died off as a result of introduced diseases, their weirs along the Nemasket were taken over by colonists until they were ordered to be removed, some as early as 1687. The first dam on the Nemasket was constructed near the site of the early weirs. The dam was used to power a forge, a gristmill, and a shovel works. The construction and maintenance of fish ladders assures that some river herring are still making it to the Assowompsett ponds, and today the run at Middleborough remains the largest in the state.

The Blackstone River, which flows almost 50 miles from the Worcester Highlands in Massachusetts into Narragansett Bay, Rhode Island, has been known by many names, including "the Seekonk, the Narragansett, the Patucket, the Neetmock, the Nipmuck, the Great" (Buckley and Nixon 2001). Anadromous fish such as salmon, shad, and river herring ran up the river and several of its major tributaries, and were found as far north as Mendon, Massachusetts. Native settlements, such as those of the Nipmucks, were located along the river in good fishing locations, commonly found just below waterfalls, where the fish tended to collect in pools. When the Colony of Rhode Island was founded, the early colonists also made use of the natural spring bounty provided by the migrating fish. Although salmon were the most prized food item, the alewives were also important, arriving just in time to be used for fertilizer during spring planting.

Regulations to Protect the Fishery

In 1709, a law to regulate the fishery was passed, which was intended to provide upstream passage to river herring so that at least some of the population would be permitted to spawn. An "Act to prevent Nuisances by Hedges, Wears, and other incumbrances obstructing the passage of Fish in Rivers" stated: "Be it enacted, That no wears, hedges, fish-garths, stakes, kiddles, or other disturbance or incumbrance shall be set, erected or made, on or across any river, to the stopping, obstructing, or straitning of the natural or usual course and passage of fish in Their seasons, or spring of the year," without appropriate permissions. If such structures were in place, they were "declared to be a common nusance and shall be demolished or pulled down, not to be again repaired or amended." These early colonial laws were the first in a series of regulations, passed over time, that were difficult to enforce. By 1727, another act was passed to make the 1709 Act more effective. A fine could be levied against those who obstructed fish passage, with half of the fee given to the poor of the towns affected by the obstruction, the other half going to the person who reported the infringement. By 1741, dam owners were required to provide sufficient water for fish passage, and by 1743 an Act was passed to "Prevent the Destruction of the Fish called Alewives and other Fish," aimed at allowing passage of migrating fish to spawning areas.

Native Americans also constructed a weir to catch river herring in what is now Pembroke, Massachusetts. The weir was taken over and used by colonists, and a public weir was still in place in 1920 (Belding 1920). Early on, any individual could harvest river herring in Pembroke, but 1741 brought the beginning of many years of regulation and management. Over time, escalating fees were charged for alewives, the season for catch was restricted, and the size of the catch was limited. Declines in the number of fish prompted a modest restocking program in 1782, when Nathaniel Cushing was allowed to take 250 herring out of Great Pond and relocate them to Furnace Pond (Cavallo n.d.). In 1742, after the establishment of several mills on Herring Brook and Indian Head Brook, the town mandated that the mills keep their gates open during the spring herring run.

The Value and Management of the Industry

During the early colonial period, the river herring fishery was a large-scale spring enterprise in New England because of the high value of the fish as fertilizer and their ability to be preserved for winter food reserves. Furthermore, as Field stated in 1914, "The special value of the alewife arises from the fact that it is one of the few fish that 'furnishes its own transportation;' coming in the early spring to the very doors of the poor people" (Field 1914, 148–49). The timing of the runs was highly predictable, so the industry afforded a dependable source of spring revenue. Furthermore, the fish were relatively easy to capture, and in some instances fish houses were built directly over a narrow section of the river or stream, or over a man-made fish passage, so that the river herring could be netted and processed directly.

As an example of the industry in Maine, both Atlantic herring from inshore marine waters and river herring, collected at the mouths of rivers during their inland migration, were collected, split, boned, dried, or smoked. Some were canned to become "sardines." It is difficult to ascertain how much of the Maine sardine industry relied on river herring (as opposed to Atlantic herring), but it is likely that they were an aspect of the endeavor. According to Ronny Peabody of the Maine Coast Sardine History Museum in Jonesport, Maine, at least some of the sardines used for the canning operations started out as so-called river herring, caught near the mouth of rivers. But Peabody points out that "a sardine is not a sardine until it is cut and put in a can" (Ronny Peabody, personal communication).

In 1896, the alewife industry reported significant harvest at the mouth of the St. Croix, and in the Dennys, Machias, Medomak, Penobscot, St. George, Pemaquid, Damariscotta, and Kennebec Rivers, as well as Casco Bay. As they were dried, each rack or "horse" of herring held 45 sticks; each stick held 30–35 herring. Fast stringers were able to assemble 800–1100 sticks per day; they were paid 30–40 cents for each 100 sticks (fig. 6.3). This spring income was an important economic factor in many towns. Fresh alewives were sold in Maine for 10 cents a pound, while smoked alewives could bring 30 cents a pound. A bushel of alewives for bait went for $3 or about 3.5 cents per fish. Preserved Maine alewives were sold by the Benson Fish Company of Chicago, which utilized the Homeport Fish Plant in Rockland to prepare and bottle them in thirty-three different sauces. A typical "old time" recipe was published in the *Bangor Daily News*, May 11, 1973: "Place smoked alewives and milk in a baking dish, dot them with butter and place sliced onions on top. Bake for 40 minutes at 350°. Serve with boiled potatoes and green salad."

Some type of pickling process usually was employed in each local curing method. A Rhode Island recipe called for placing alewives in the "strongest

6.3. Racks of herring prepared for smoking (NOAA Fisheries)

possible brine" for a day or two, after which they were turned over and stirred every two days. Subsequently, they were packed in barrels containing two bushels of salt per barrel. A method, used by George M. Besse, of Wareham, Massachusetts, involved covering the fish with salt on the packing room floor, and then placing them directly in barrels that held two inches of salt and two inches of water. For pickling, the fish were stacked in layers that alternated with salt and then the barrel was half filled with water, remaining so for one month. If the fish were headed to market, they were packed dry with plenty of salt; if they were to be smoked, they were strung through the eye sockets and brought to the smokehouse (Belding 1920).

River herring are oily, bony fish that many of us have probably not had the opportunity to savor. As early as the 1800s, fresh, salted, smoked, or pickled river herring dropped out of favor as a food source. With the availability of ice and the advent of refrigeration, many species of fresh-caught fish could be easily transported and became much more available to consumers. But river herring remained a valuable commodity, primarily as bait for commercial fisheries for cod, haddock, and lobster and also in uses similar to menhaden: fish oil, fish meal, fertilizer, and animal food. The method of harvest varied with the type of run, the size of the river or stream, and the optimal harvest location—the river or stream itself, the mouth of the river, in a harbor, or in the headwater lakes and ponds. Weirs were often employed to corral the fish, seine nets were used near shore to collect them, gill nets were used to enmesh them, and simple dip nets were used to scoop them out of the water. In the Mid- and South Atlantic, pound nets were sometimes employed. This device is constructed from a series of net-bound enclosures, which direct the fish into a trap called a "crib" where they could be netted from a boat.

By the late 1800s, the bulk of the commercial alewife fishery was located in the Mid-Atlantic states, with Virginia, Maryland, and North Carolina furnishing the most fish. In the Northeast, the fishery was not as extensive, but it was still an important enterprise in Maritime Canada, Maine, Massachusetts, Rhode Island, Connecticut, New York, and New Jersey, where fish were caught inland and also mingled among the menhaden and mackerel in weirs and pound nets along the coast.

The inland river herring fishery in Massachusetts has undergone a major transformation. The fishery was initially open to all inhabitants of a town or village, but after time, regulatory actions were imposed as a stopgap means of preserving the resource. Local herring committees were formed, and each town or cluster of towns would hire a herring warden. The committees were autonomous and managed the fisheries as they thought best. Herring leases were sold or auctioned to individuals or companies, and in many cases, the entire fishery was taken over by the town and municipally harvested fish could be purchased by residents. Decisions were made about management of individual runs; in some cases, limitations on harvest days and amounts were deemed sufficient to maintain the industry. But local oversight did not necessarily mean that the runs were well managed. As early as 1730, the citizens of Plymouth were restricted to taking only four barrels of alewives each day, but by 1815 they were fortunate if they could net two hundred fish. In other cases, the decreases in harvest or loss of stocks from runs were met by attempts to replenish fish by restocking programs.

In the 1880s, "Herring Time," as it was called, was still very much part of the pulse of coastal communities. In the village of Bournedale, the annual run was anticipated as soon as the ice broke up in Herring Pond and its waters would flow downstream and hit tidewaters. The local herring fishermen would spend each winter trimming sections of spruce to make net handles, steaming sections of ash and white oak so they could be bent into net bows, and then drilling small holes around the bows so that they could be laced with twine. An investment of 75 cents was required for the handle and bow and another 50 cents for the net. The total of $1.25 was more than three quarters of a man's daily wage at that time, but well worth the investment if the net would bring in alewives. A local historian wrote, "He who brings in the first shining alewife is proclaimed and honored in the village; the word of his triumph quickly passes from house to house" (Jacobs 1996, 27). Until April 15 of each year, the fishery was open to all, but after that date, the right to harvest alewives was restricted to the person or company who won the exclusive right to harvest at auction, usually costing that person a few hundred dollars. Although fishing rights were sold to companies, Bournedale, like other coastal New England villages, maintained a system in which village inhabitants could still claim, from the company, one barrel of herring per taxpayer or voter. In addition, the head of every family of Herring Pond Indians was entitled to one free barrel. Eventually, fishing was restricted to a few days each week, but poaching, under the cover of darkness, was a frequent occurrence as alewives were caught, packed in barrels, and carted over back roads to waiting ships in need of bait. If there was an excess of fresh fish, i.e., more than was needed for bait, they were salted, sent to the West Indies, and fed to slaves on the sugar plantations; in trade, the herring suppliers received blackstrap molasses, which was used to make rum. A similar industry was important in Connecticut. River herring, primarily from Wethersfield and Rocky Hill, were shipped in barrels to feed slaves on the sugar cane plantations of the Caribbean in return for molasses. World War II saw the demise of the Wethersfield fishery, while Rocky Hill operated until the early 1970s (Connecticut River Coordinator's Office 2004).

Not all rivers and streams are restricted to town or even state boundaries. The Blackstone River supported anadromous fisheries in both Massachusetts and Rhode Island until salmon disappeared in the 1700s, followed by shad and herring, which were commercially harvested until the end of the 1800s. Massachusetts towns often petitioned the General Assembly in Rhode Island, contending that Rhode Island industries erected obstructions to fish passage that hampered the fishing rights of Bay State residents (Buckley and Nixon 2001).

The Mattapoisett River was originally within the town of Rochester, Massachusetts, but Rochester was subdivided into three separate towns: Rochester, Marion, and Mattapoisett, with three separate governance systems. However the herring fishery remained under joint control, with any profits and losses divided up to reflect the amount of taxable property contained within each town.

A U.S. Bureau of Fisheries Report, dated 1912, describes seventy-four river herring fisheries in Massachusetts, including Nantucket and Martha's Vineyard, in rivers, streams, and coastal ponds. The towns, listed alphabetically, ranged from Acushnet to Weymouth and provided a combined commercial revenue of $45,000 in 1908, translating to over one million 2010 dollars.

In a 1920 report, Belding described the system for assigning thirteen privileges to seine herring along the Taunton River, distributed to the towns of Middleborough, Raynham, Taunton, Berkley, Dighton, Freetown, Somerset, and Fall River. The privileges were sold outright or auctioned to the highest bidder. The person who purchased the privilege was allowed to seine for river herring anywhere along the Taunton and its tributaries. When Belding wrote his report, the alewife fishery was already in serious decline, as exemplified by the lack of interest and the decreased cost in seining privileges. "The Dighton privileges which formerly sold for $400 to $500 now sell for $10 to $20. In 1913 the three Taunton privileges which in 1899 cost $45 were sold for $10 apiece. The city of Fall River in 1880 sold its privilege for $103; in 1884 for $50; in 1906 for $7.50; and in 1909 for $21. Since 1909 the privilege has not been sold" (Belding 1920, 113).

Current Moratorium

Once an important business that subsidized the expenses of local towns, helped to pay salaries of town employees, and supported local schools, the river herring fishery eventually took a back seat to industrial and farming interests. At the turn of the twenty-first century, as towns and states began to notice the very low, or sometimes absent numbers of river herring, confirmed by regional stock assessments and local declining herring count data, many areas began severely limiting, and even closing their commercial and recreational fisheries. Some states began to ban harvest, with Connecticut, Massachusetts, Rhode Island, and North Carolina leading the way. Although not required to do so because of their tribal status, even the Aquinnah Wampanoags closed down the commercial end of their fishery, coincident with the Massachusetts river herring fishing moratorium.

Major federal harvest restrictions were placed on the river herring fishery in 2009, when the Atlantic States Marine Fisheries Commission (ASMFC) revised the Interstate Fishery Management Plan for Shad and River Herring by passing Amendment 2. All river herring fisheries, whether commercial or recreational, would be shut down by January 1, 2012, unless a town or state developed a sustainable management plan, subject to review and approval by the ASMFC. While many states are still abiding by the moratorium, Maine has managed to keep its river herring fishery open for business. The fishery is managed so that a portion of the adults will be able to spawn by limiting the number of runs in which fish can be harvested and by closing the fishery three days per week to allow escapement, i.e., the opportunity for a subset of fish to make it to spawning grounds. Although it is not a major fishery, the potential for closure was predicted to lead to significant economic losses and served as the impetus for the formation of Alewife Harvesters of Maine, an organization that identifies itself on its web site as "a group of alewife harvesters, conservation commissioners, biologists, and concerned citizens who have joined together to conserve alewives, and to preserve the river-fishing heritage of Maine." Approximately eighteen runs are managed by local municipalities, while the Maine Department of Natural Resources also has the right to lease fishing privileges on runs that are not managed locally.

On a wet, gray, May day, my husband and I traveled to the Gulf of Maine Research Institute in Portland to meet Jason Stockwell. I had been to the Stockwell lab prior to this visit to observe how otoliths were removed from river herring and prepared for microscopy. This trip had a different focus: it was planned to observe live river herring on their spring migration. As we left the lab, a very heavy mist kept our windshield wipers squeaking back and forth for the entire trip to the fishway at Damariscotta Mills, our final destination. A springtime tourist attraction, the stone and mortar fishway has been undergoing repairs and restorations for a number of years. It is situated on the Damariscotta River, on a steep part of the 42-foot rise between the ocean and spawning grounds on Damariscotta Lake. With bends, turns, and resting pools, it appears to be a well-planned fishway; netting covered narrow and vulnerable sections to provide a barrier against avian predators. The only aspect of the run that seemed out of place was a large chute (fig. 6.4). Stockwell explained

6.4. Alewife chute for commercial harvest, Damariscotta, Maine (photo: Barbara Brennessel)

how the operation works. On certain designated days each week, the chute gets connected to the fishway and, under the watchful eye of the local fish agent, fish are collected and funneled into containers belonging to alewife harvesters whose customers await their lobster bait. There were no passing fish while we were at Damariscotta Mills on that dreary spring day, a disappointment to us and also to the hovering gulls, but we nevertheless rewarded ourselves on the way back to Portland with a terrific lunch at Red's Eats, a trailer-sized lobster shack. It seems ironic that Red's lobster roll is top-rated but it may have been a Damariscotta alewife that served as bait for our lunch.

CHAPTER 7
Dam Yankees

Almost from the time they arrived in New England, European settlers began to engineer the land and waterways. They cut down trees to build homes and cleared land for farming; they diverted and filled rivers and streams to reclaim land, and they constructed dams to control flooding and impound water. Many of the dams served to redirect water flow over mill wheels and turbines in order to harness power.

Powering Up and Changing the Flow

Dams can range from something as simple as a small earthen mound or pile of rocks across a few feet of water to some of the massive structures across large waterways such as the Connecticut River. Many of them go unnoticed by the general public because they are on private lands and may not be regulated or inspected by state or federal agencies. Every river and almost every stream in the Northeast has been altered by the construction of one or more dams.

Some of the earliest dams, small structures compared to their modern counterparts, were constructed for the purpose of impounding water and then diverting it over a large wheel that could be used to grind corn; the

7.1. Jenney Grist Mill, Plymouth, Mass. (photo: Barbara Brennessel)

Mill at Stony Brook in Brewster, the Jenney Grist Mill in Plymouth, and the Gilbert Stuart Mill in Saunderstown, Rhode Island, are good examples. As Plymouth developed from a small colony to a thriving town, Town Brook remained an important asset to local citizens. Over the years, eight dams were constructed to power mills and factories (Belding 1920). As early as 1632, a dam powered the mill built by Stephen Dean to grind corn. The mill was rebuilt several times, once in 1636 by John Jenney, for whom it is still named. When it was rebuilt by Charles Stockbridge, it had been in continuous operation for 213 years; the reproduction mill at the historic site still operates . . . as a tourist attraction (fig. 7.1).

In 1683, the Plymouth town fathers realized that the Jenney Grist Mill was impeding the passage of herring to spawning grounds and came to an agreement with Stockbridge that something must be done to allow the fish to circumvent the obstruction, so a water course for herring was constructed at the mill. Over the years, river herring lost the battle with industrial interests; a fish bypass was no longer deemed important and Town Brook was used as a dumping ground for industrial waste and sewerage from private residences.

Although many laws were enacted to preserve the coastal herring runs in Massachusetts (which included Maine) and Rhode Island Colonies during the early colonial period, the laws were ineffective or not enforced. In 1735, "An Act to Prevent the Destruction of the Fish called Alewives" required Massachusetts mill dam owners to provide passage for migrating alewives. In 1741 the law was further clarified, to state that dam owners must provide, and also maintain, fish passage between April 1 and May 31 of each year and that they allow sufficient water flow for the young to emigrate in the fall. In the 1700s, similar laws were passed in Rhode Island, which witnessed the construction of a multitude of dams and the assignment of water privileges during this period. By the early 1800s, there was approximately one dam for each mile of the Blackstone River (Buckley and Nixon 2001).

After the colonial era, dams continued to be constructed to store fresh water, to generate power for industrial applications such as the old axe factory in Bournedale and the Woolen Mill on the Parker River in Essex, Massachusetts. Water privileges were up for grabs, and farmers constructed water sluices for irrigation or to harvest cranberries. Some of the early dams in Maine were built by logging companies to facilitate the log drives, which used the rivers as highways to deliver felled and trimmed trees to wood and paper mills. The needs of the growing country and the advent of the industrial revolution continued to change the course and flow of rivers throughout the Northeast and to overshadow the importance of the historical anadromous fisheries, which were no longer considered to be lucrative industries. Many of the laws that protected fish passage were relaxed to accommodate pressures from the growing industries, which relied predominantly on hydropower.

More modern twentieth-century dams in the Northeast are much larger structures and were constructed for flood control and to store fresh water in reservoirs as well as to generate power. On the Connecticut River, the Holyoke Dam initially served as the power supply for a number of industries, including textile mills, while today it generates hydroelectric power that is sold to the national grid.

No Place to Run

Whatever the reason for constructing a dam, its presence will have profound effects on the river or stream on which it is built. Dams are not only barriers to upstream fish passage, they also pose threats to downstream migration of spent adults and, later in the year, to juveniles as they make their journey to the ocean. Fish may suffer mortality in the turbulence of spills over large dams or impact rocks or man-made structures at the base of the dam. Dams change water pressure, an event that can be fatal to juveniles. They may induce a delay in migration and provide a magnet for predators, for example, if fish concentrate above or below the dam in quiet pools or slow-moving water while migrating upstream or taking the plunge downstream. Some downstream migrating fish may be trapped in turbines of energy generating plants where they may be converted into minced herring or become so stressed that they easily suffer from disease or depredation.

In addition to interfering with the passage of migrating fish, dams have far-reaching effects on river ecology. The impounded water floods former shoreline locations and can change dry upland habitats to wet shoreline habitats. Near the coastline, dams can restrict tidal flow and tidal flushing, thus changing water composition and decreasing brackish water and salt marsh habitat. By blocking natural water flow and mixing, dams can affect water temperatures and thus have profound impacts on living organisms and natural temperature-dependent processes and conditions, such as oxygen composition. Added to these consequences, dams also serve as physical barriers, which retain sediments and concentrate pollutants. This latter fact is an important finding that must be seriously addressed in any attempts at dam removal.

But our primary concern is fish passage or lack thereof. Rhode Island is a case in point. The construction of over five hundred dams across streams and rivers in the tiny state of Rhode Island has altered the flow of water from source to coast. Dams on the Taunton, Blackstone, Pawtuxet, and Ten Mile Rivers have impaired at least forty-five herring runs that were known to exist in the Narragansett Bay Watershed. Today, herring run in eighteen Rhode Island streams but the Rhode Island Department of Environmental Management has identified about forty-one streams that would have the potential to support herring runs if they were properly restored (Rhode Island Restoration Website).

The same is true for other northeast states with historic herring runs. The exact number of dams varies according to the source and according to the size and classification of the dam. For comparative purposes, I consulted the Report Card for American Infrastructure collated by the American Society of Civil Engineers, a group that assesses the condition of roadways, dams, and bridges throughout the country. The Society lists 1,630 dams in Massachusetts, although I have seen estimates as high as 3,000 dams on 10,000 miles of rivers and streams; 1,187 dams in Connecticut, with other sources estimating as many as 9,000 dams along 32,000 miles of rivers and streams, 643 dams in Rhode Island, 3,073 dams in New Hampshire, 831 dams in Maine and 5,089 in New York. To be clear, not all of the dams are located on waterways utilized by anadromous fish, but a large number of them may be in vulnerable migration waterways.

Historically, there were many more dams than exist today, but many have been removed and others are slated for removal. In addition to the fact that many are no longer functional, some pose safety issues or were removed as part of the planning and management for restoration of particular waterways. The Whittenton Mills Dam on the Taunton River in Massachusetts was in precarious condition as a result of an October nor'easter that hit southern New England in 2005. Town officials issued evacuation orders as rising waters threatened to breach the dam, flood the area downstream, and damage the homes of two thousand people. The dam was constructed in 1832 to power the adjacent mills, but over the years, it ceased to serve the function of a power source, and eventually became part of a parcel that was owned by a real estate development company. After the storm, the dam received an emergency, temporary fix to stabilize the decaying structure, but for this, and similar decrepit dams, it is only a matter of time before another crisis occurs. Proposals to remove the Whittenton Mills dam, along with others on the same stretch of river, would lead to the opening of over forty miles of the Taunton as it winds to Narragansett Bay, and offer an unimpeded inland route for spawning river herring.

Engineering the Landscape

Although dams are the most serious impediment to fish passage, other construction projects have had negative impacts on the inland movement of river herring. Preliminary genetic evidence suggests that the Erie Canal may have introduced invasive alewives to the Great Lakes via the Hudson/Mohawk River system (Ihssen, Martin, and Rogers 1992), and a similar introduction may have occurred through the Welland Canal, between Lake Ontario and Lake Erie, built in 1824 to circumvent Niagara Falls and to connect the Great Lakes to the St. Lawrence Seaway. Invasive alewives, whose path to the Great

Lakes was made possible by the canals, have formed permanent landlocked populations that provide unique challenges to fisheries managers.

In southern New England, the Cape Cod Canal connects Buzzards Bay and Cape Cod Bay. The idea for the canal was first suggested in 1623 by Miles Standish, military leader of Plymouth Colony, as a way to connect Plymouth Colony with an important post at Aptuxcet. Here, the Pilgrims had an opportunity to barter with Dutch traders arriving from New Amsterdam as well as Native tribes from coastal Rhode Island and Connecticut. Without a canal, the Pilgrims were forced to portage their vessels between the Manomet and Scusset Rivers. Despite many aborted attempts to raise funds, it wasn't until 1914 that the first canal was built by a private company, which charged each vessel a fee for passage. This canal, which was only 100 feet (30 meters) wide and considerably shallower, was eventually purchased by the U.S. government as a public waterway. During the Great Depression, the U.S. Army Corps of Engineers hired over a thousand men and widened it to 480 feet, thus making it the widest sea-level canal in the world.

After the construction of the canal, the southeast coast of Massachusetts was re-sculpted. Vessels can bypass 135 miles of open and sometimes treacherous ocean, site of many shipwrecks and fraught with menacing currents in Vineyard Sound, around the shifting shape of the narrow Monomoy peninsula. Of the 17.4-mile (28-kilometer) length of the project, 7 miles (11 kilometers) traverse the landmass of Cape Cod. The canal subsumed the Monument (originally Monomet) River and changed the path of its tributary, the Herring River, which connects the canal to Herring Pond, the freshwater spawning area for river herring. In essence, it cut off the natural flow of water from the inland ponds to Buzzards Bay. In making a safe passage for large vessels, the Cape Cod Canal created problems for small fish. In 1936, to remedy the loss of access to spawning habitat, the U.S. Army Corps of Engineers constructed a herring run consisting of a series of concrete pools and fish ladders, connected with culverts and PVC pipe. The start of the run is in the Cape Cod Canal, near the Canal Visitor Center, then up a concrete fish ladder between the canal and Route 6, known locally as "Scenic Highway." The run passes under Scenic Highway to a resting pool at the Herring Run Motel. This is the former site of Benoit's cottages, where the river herring could be netted if one had the proper permit. In the 1960s, a $75 permit allowed the permit holder to scoop out a 30-gallon trashcan full of fish (fig. 7.2). Eventually, the number of the fish decreased to a point in

7.2. Commercial fishing at the Herring Run Motel swimming pool Bournedale, Mass., ca. 1950 (courtesy Bourne Archives)

which town employees became responsible for harvesting herring from the run and individuals were required to pay $25 to fill their bait boxes with 15–25 fish. When Robert Finch visited the pool at the Herring Run Motel in preparation for his book, *Special Places on Cape Cod,* he saw a sign which stated, "No Lifeguard on Duty—Swim at Own Risk—No Diving." He further noted, "The pool was black with swarming alewives who didn't seem worried about the lack of lifeguard and certainly didn't seem to be trying to dive" (Finch 2003, 29).

Other man-made structures that can take their toll on natural processes in the river herring life cycle are culverts—less expensive alternatives to bridges in narrow sections of streams and creeks. Culverts are tunnels placed under roadways or embankments; they may be constructed in such as way as to narrow the natural width of a stream and, in addition, they may cause migration delays. Sluice gates, such as those that were constructed in the early 1900s on the Herring River in Wellfleet as a mosquito control measure, also present a number of challenges to upstream migrating fish. Sluice gates prevent tidal flow, alter the characteristics of rivers, streams, and estuaries, and pose a physical barrier to migration. Natural dams, such as those built by beavers, are constructed out of logs, stones, and mud. North of Boston, beavers have impeded water flow in a number of areas such as Alewife Brook, which connects the Essex River to spawning grounds on Chebacco Lake. In Essex, the problematic beaver dams are removed by hand to allow upstream passage of spawners and downstream passage of adults as well as young of the year. Because beavers use intricate architecture and are prolific builders, their obstructive homes must be constantly monitored in areas where river herring attempt to pass.

Other types of natural dams are those formed by fallen timbers and vegetative debris. Some of these impediments may actually serve an important function by affording areas where fish are protected from prey or providing substrates to which eggs may adhere, but when the impediments build up to a point that they prevent passage or cause stress, they can do more harm than good.

Although natural processes contribute to the erection of barriers to fish migration, most of the obstructions are the result of human activity. In addition to structures such as dams, canals, sluices, and culverts, humans have constructed other types of structures as part of water projects for residential, commercial, agricultural, and recreational use, in which massive amounts of water are withdrawn from watersheds for purposes of supplying water to homes and businesses, irrigation, and snow making. Such water withdrawals can alter river and stream characteristics such as temperature, current flow, and substrate. If not equipped with proper screens, water intakes may trap fish and cause mortality. Changes in land use have led to dredging, filling, channelization, and draining of wetlands. It seems that throughout the history of the Northeast, there is no limit to our inventiveness in altering natural water flow to suit our needs. As the importance of the proper functioning of natural ecosystems becomes more and more apparent, the types of projects that have negative impacts on river herring migration are diminishing, and projects to rectify some of the problems caused by inappropriate historic blunders are being initiated.

CHAPTER 8
Bogged Down

In the fall of 2010, a considerable number of young of the year river herring became trapped in a cranberry bog in Falmouth, Massachusetts. Apparently, the cranberry grower did not install the proper type of net or mesh screen, or the screen did not function properly as a barrier to block the passage of the young herring into the bog from Wing Pond. During this crisis, some observers estimated that half a million herring were trapped and a considerable number were subsequently macerated by a water pump that was used to flood the bog just prior to the cranberry harvest.

Before Europeans set foot on our shores, cranberries and river herring coexisted in southern New England, and they were both critical to the survival of the first settlers: river herring for sustenance and to fertilize corn crops, and cranberries to prevent scurvy. The important nature of both resources is depicted on the seal of the Middleborough Historical Commission (fig. 8.1), where river herring and cranberry vines are placed in prominent locations to indicate their equal importance to the economy of the town.

Before the days of commercial cranberry culture, cranberries were naturally found in sandy bogs and shallow freshwater marshes. These natural bogs are still scattered about in small patches in parts of Cape Cod in serene swales protected from the wind, much as they existed before the berries became an

8.1. Middleborough Historical Commission Seal (designed by Janet Griffith; logo is provided courtesy of the Middleborough Historical Commission)

agricultural staple. After cranberry growing went commercial and evolved into a lucrative industry, New England towns were able to derive more tax income from cranberries than from fishing rights. The river herring fishery also came to be considered of less economic importance when cranberry farming became profitable because the number of fish had been declining for decades. A Plymouth cranberry grower was known to purchase fishing rights on Fresh Pond Stream so that he could manage the stream, not to harvest river herring, but to optimize the yields from his cranberry farm.

Cranberries, known as lingonberries or English mossberries in Europe, can be found in northern latitudes worldwide; however, the cranberry that we prize for its red fruit with high anti-oxidant value (it has one of the highest oxygen radical absorbance capacities of all common foods) originated exclusively in North America. The American cranberry (*Vaccinium macrocarpon*) is used in snack foods, juices, and juice drinks, and as a prime ingredient in sauces that are served with meat and fowl. One rarely has a Thanksgiving dinner without turkey and its accompaniment, cranberry sauce. Found from eastern Canada south to the highlands of North Carolina, cranberry plants grow in many regions which historically supported large herring runs. Called "sassamanesh" by the Wampanoags, they were an important constituent of the Native American diet, eaten raw, and also mashed with dried meat and fish and formed into cakes that were air-dried. This preserved food, known as "pemmican," was high in protein and vitamin C and served the Natives well as a travel provision. The berries were also used as a source of red dye and incorporated into many Native medicines.

The Natives picked cranberries by hand from low-lying bushes in bogs where they grew naturally: special types of freshwater marshy areas, in peat, a type of acidic soil that also contains gravel, sand, and clay. They introduced the colonists to cranberries, just as they had done with alewives, and before long, the berry became part of the colonial diet.

In 1813, Henry Hall, a Revolutionary War veteran who lived in East Dennis, on Cape Cod, devised a way to plant and cultivate cranberries. He capitalized on growing them for use as a scurvy preventative, selling them to seafaring and trading companies, and at the same time, his cranberry-growing operation served as a model for commercial production. By the mid-1800s, several Cape Codders followed Hall's lead, including growers in Harwich, who dug canals between Hinckley's Pond and the Herring River and built dams and dikes to control water needed for irrigation of the bogs. Some went so far as to create cranberry bogs where none existed. In 1891, the Swift brothers transformed an area of Falmouth into a cranberry bog by altering the Coonamessett River, using techniques such as straightening specific sections of it, and constructing dams, channels, and dikes. Similar methods were employed wherever cranberries were commercially grown. The diverted water was used to irrigate the crop and also to flood the bogs at harvest time. Methods to foster cranberry agriculture were usually problematic in terms of maintaining river herring runs and upstream spawning habitat.

The Marstons Mills herring run, which can be traced from Nantucket Sound northward to the Middle Pond spawning area, was virtually nonfunctional by the end of the 1800s. Construction of grist mills along the river by the Marston family certainly took its toll on the number of the fish that were caught in the run each spring. There was a steady decline in the value of fishing rights on the river, which hit an all-time low of $35 for a five-year lease in 1879. But it was not only the mills that contributed to this sorry state of the fishery, because farmers had become aware of the value of the land and ponds in the Marstons Mills watershed for cranberry agriculture. Abel Denison Makepeace, the Cranberry King, was one such farmer. The company he founded, A. D. Makepeace, remains the largest cranberry grower in Massachusetts. In 1875, Makepeace began the legacy of cranberry growing in

Marstons Mills by diverting water from the Marstons Mills River by constructing ditches and canals. If the exploitation of the river herring fishery and the obstructions presented at the mills had not already destroyed the herring run in Marstons Mills, the cranberry farms certainly did. Tension between the river herring fishermen and the cranberry farmers was noted in 1920 when David Belding wrote, "The interests of the fishery and cranberry industry are diametrically opposed. In the welfare of the latter the course of the natural stream is changed, channels are made, water diverted by ditching, and dams erected for reservoirs or for flooding the bogs" (Belding 1920, 50).

There are two main types of cranberry-growing operations. Most bogs are dry bogs, physically separated from streams and ponds by berms or roadways. Water, needed for irrigation or harvesting, is pumped into the bogs from a nearby water source, and, to prevent fish kills, water recovery ponds hold the used water for specified periods of time before it is released back into the main water supply. The "hold" is to allow decomposition of some of the chemicals used in the agricultural operations. The second type of cranberry farming, constituting a much smaller percentage of commercial bogs, is called the "flow through" bog, grandfathered to operate in Massachusetts in this manner prior to the passage of the 1996 MA Rivers Protection Act. These bogs, as their name implies, have a stream or watercourse flowing directly through them; dams hold back the water until it is needed, and thus this type of bog causes more problems in terms of concentrating agricultural chemicals and obstructing fish passage.

Typically, the cranberries grow during spring and summer when the bogs often require irrigation; the crop is harvested in fall, when the berries ripen. A small percentage of cranberry crops are dry harvested, using mechanized raking devices that are attached to conveyor belts, delivering the berries into collection bags. However, most of the crops are harvested by wet picking, i.e., flooding the bogs, typically from September though November, coinciding with the emigration of the young of the year river herring. The berries, removed from the vines with a harvester, float to the surface and are corralled into one area where they can be scooped up and packed. In order to flood the bogs, water from adjacent rivers, streams, or ponds must be diverted or, in some cases, it is released from an impounded area, usually within an earthen dam containing a sluice gate. When the water is pumped into the bog to float the cranberries, the intake pumps must be properly screened to prevent river herring from being sucked into the bog and subsequently trapped. In addition, the creation of artificial waterways such as ditches and canals not only prevents the mature river herring from reaching spawning areas in the springtime; these practices may also have a negative impact on young fish. For example, diverting water or opening a sluice gate has the potential to dry down headwater ponds and prevent sufficient flow to allow the young to emigrate.

Although many modern growers are attempting to employ Integrated Pest Management techniques, many cranberry farms still use a considerable number of herbicides, pesticides, and insecticides, which can be fatal to fish. There are many documented cases of fish kills resulting from agricultural chemicals used by cranberry growers.

In the late 1800s and early 1900s, there were many conflicts between alewife harvesters, who were already seeing marked declines in their catch, and cranberry growers, whose water usage decreased the size of the runs even further. The popular antioxidant properties of cranberries and the development of novel uses for them in food products has destined this crop to be a major economic driving force in areas where it can be cultivated. Thus, many efforts have been under way to restore the harmony that once existed between the little red berry and river herring. Eventually the growers were mandated to provide safe passage for anadromous fish. The MA 1996 Rivers Protection Act directly addressed fish passage with respect to cranberry agriculture. Abandoned and nonfunctioning cranberry bogs are particularly problematic, so even owners of decommissioned bogs are required to construct bypass canals to allow fish passage. The work of the Eel River Headwaters Restoration Project transformed a decommissioned cranberry bog in Plymouth into wetland habitat that included a functioning headwater stream, thus providing for fish passage in the system. The Cape Cod Cranberry Growers Association has published an *Advisory for Anadromous Fish Passage*, which informs growers that impeding passage of anadromous fish is illegal, and reminds them to avoid holding water and reducing flow in the spring (which would hinder upstream migration to spawning areas), to construct fish ladders where passage may be a problem, and to ensure sufficient flow in the fall so that juvenile emigration is not prevented or delayed (Cape Cod Cranberry Growers 2004). Growers are also required to have the appropriate screens installed on the intake side of water pumps prior to any bog flooding operations and to harvest their crops immediately. If the bog remains flooded for an extended period in the fall, a week or more, the bog owner must consult with a local herring official or the MA Division of Marine Fisheries. If cranberry growers follow these

Best Management Practices, perhaps the peaceful coexistence between the cranberry and the river herring that occurred prior to the arrival of European settlers on New England shores can be forged anew.

Habitat Alteration

As well as overfishing, and reconfiguring rivers by damming and rechanneling waterways, other human activities may be responsible for declining river herring populations, including development, farming, and industry. Reports of river herring spawning distribution along an urban gradient (Limburg and Schmidt 1990) implicate human use of the landscape as a factor that impacts river herring reproduction. So what are humans doing? How are river herring affected? The answers to these questions are not straightforward. It may be a combination of activities that are collectively leaving their mark. Early on, before any serious attempts to monitor and limit what was being dumped in American waters, raw sewage often found its way into rivers. In addition, the rivers and their watersheds were virtual soups of chemical by-products, trade wastes such as dyes, acids, and alkalies, produced by industries such as woolen mills, dye factories, and iron works (Belding 1920). As attention was paid to the elimination of sewage and chemical dumping, other habitat alterations went unnoticed. Some examples are provided to illustrate the seriousness of problems that were initially unobserved.

Acidification: Acid rain was recognized as a detriment to ecosystems in the 1970s and 1980s, spurring much research and resulting regulations to limit the amounts of acid-containing pollutants from power plants. Most agree that some improvements have resulted from the attention paid to this environmental problem. However, rivers, streams, and lakes may not experience acidification on a regular basis, but if it is occurring episodically, at critically sensitive periods, such as during larval and juvenile development, it could mean disaster for river herring stocks. As determined in controlled laboratory studies, pH values below 6.0 may be particularly harmful to blueback herring embryos and yolk sac larvae (Klauda and Palmer 1987; Klauda, Palmer, and Lenkevich 1987). In addition to the effects of pH alone, acid waters may enhance the deleterious effects of certain chemicals. For example, if aluminum, iron, or mercury is present, these elements will have higher toxicity in an acid environment. Therefore, it is no surprise that episodic acidification can be a problem for river herring, especially during early developmental stages, but acidification in areas where industrial wastes contain heavy metals such as aluminum, iron, and mercury can be particularly problematic.

Nitrogen Loading: Human activities are responsible for pouring nitrogen into our watersheds. Nitrogen is a limiting nutrient, and therefore important in supporting aquatic life. However, too much nitrogen, flowing into systems as a result of agricultural and fertilizer runoff as well as human and animal waste from point and non-point sources, can cause eutrophication, leading to dramatic changes in aquatic ecosystems. Increased nitrogen fosters algal blooms, which can lead to depletion of oxygen and subsequent fish kills. Some of the algae that can proliferate under eutrophic conditions may produce toxins, an additional problem if they target fish.

Dredging: Waterways are often dredged to provide access for commercial and recreational vessels, and as a result of the process, the local ecosystem may be altered in a number of ways. Increasing siltation may change the nature of the substrate, possibly impacting spawning. Increases in turbidity may change feeding ecology, and contaminants, which were trapped in sediments, may become suspended, leading to toxic repercussions. In addition to these potential problems, if the timing of dredging coincides with migration to spawning grounds, fish, eggs, or larvae may be physically harmed.

Modern Industrial Waste: A small waterway in Rhode Island provides an interesting example of a waste product from modern aviation that can impact a river herring run. T. F. Green Airport, outside of Providence, uses glycol de-icing compounds in the winter to prepare planes for departure. The glycols were making their way into Buckeye Brook, a natural fish run that allows passage of river herring from Narragansett Bay to a spawning area in Warwick Pond. These chemicals are toxic to fishes and other aquatic life because they are metabolized in the environment by a process known as biodegradation by aquatic bacteria, which use lots of oxygen to degrade the chemicals, resulting in a reduction of oxygen available to aquatic wildlife. Like other causes of decreased oxygen availability, this type of biodegradation can also result in fish kills. Since the time that the Rhode Island pollution was confirmed, the airport has employed a mitigation method in which the de-icing chemicals are captured before they reach the watershed.

Water Withdrawal, Thermal Discharge, Impingement and Entrapment: Power plants along the East Coast that withdraw water near sensitive river herring habitat include energy generating plants at Seabrook nuclear plant in New

Hampshire, the Pilgrim nuclear plant in Plymouth, and the Brayton Point Station, located on Narragansett Bay in Somerset, Massachusetts, the largest steam-electric generator in the Northeast. Withdrawal of water to use for cooling and return of heated water into rivers and embayments can be detrimental to all forms of aquatic life. In order to cool a power plant, water is removed from the nearby water source; this in itself may cause problems with fish populations if eggs and larvae become entrained on the water intake structures or if fish get trapped in the screening system. But the most problematic ecological aspect of these plants is the heating of water and its immediate return. For example, at Brayton Point, 1.45 billion gallons per day of hot water is discharged into Mount Hope Bay. Although not directly linked to the declining numbers of many types of fishes in the bay, there is potential for heating of the bay waters to temperatures which are problematic for many species. A model was developed for the flow of the thermal effluent at Brayton Point, which resulted in the setting of a maximum temperature for discharge water and a thermal mixing zone to limit the area for thermal effects, attempts to minimize the impact on aquatic life.

Clean Water

Despite the potential problems posed by chemicals and other pollutants, most observers concur that our rivers have been cleaned up considerably. The Federal Water Pollution Control Act of 1972, better known as the "Clean Waters Act," has led to major, measurable improvements in many waterways. Although it is acknowledged that some rivers are still impaired, and may never be restored to pristine historic conditions, it may be possible to swim and fish in rivers that were once loaded with human and industrial waste. Our improved water quality does not seem to be benefiting river herring, however. As our rivers are getting cleaner, river herring populations remain on a downward trajectory. It seems obvious that other factors are at play.

The Dynamic Food Web

One reason that has been posited for river herring decline in the Northeast is the increase in the populations of river herring predators. There are three species that serve as examples of the consequences of managing or protecting predators that prey on river herring: striped bass, grey seals, and cormorants.

Striped Bass

In the early days of the Plymouth Colony, the striped bass (*Morone saxatallis*) fishery was a valued commercial industry. According to the MA DMF, one of the very first free (public) schools in the colonies was established in 1670 with income from this fish, which were packed in barrels and shipped to England in exchange for currency and building materials for the school. Stripers have remained an important target species for commercial as well as recreational fishermen; most fishermen are aware that striped bass are partial to a meal of river herring. Anglers know that live-lining with forage fish such as menhaden, mackerel, or river herring is one of the best methods to catch striped bass. During springtime, as soon as the river herring begin to congregate in estuaries, at the mouths of rivers and streams, striped bass fishermen will not be far behind. When it was legal to do so, striped bass fishermen might be seen at the local herring run before they made their way to their favorite fishing spots. They would collect a few dozen river herring and try to keep them alive, perhaps in a live bait well on their boats. The herring would usually be hooked transversely across the nose and let out on the fishing line.

Striped bass populations have seesawed during the past one hundred years with periods of scarcity followed by periods of partial recovery. For example, from 1897 until the beginning of the 1920s there was a thirty-year dearth of stripers north of Boston, but then followed a period of increased relative abundance. A steady decline in the numbers of striped bass began in the 1970s, leading Congress to amend the Anadromous Fish Act, which has since been amended several additional times. The amendments initially provided minimum size limits and were designed to protect females in order to assure that they would be able to spawn at least once. Over the years, daily catch limits were modified for both commercial and recreational fisheries. By 1995, the Atlantic States Marine Fisheries Commission deemed that the striped bass population along the Atlantic Coast was "recovered," but numbers dipped slightly in the first decade of the twenty-first century, so management efforts continue to be in place in attempts to prevent another striper crash.

At the same time that the striper population has bounced back, one of their preferred prey items, the menhaden, has been in short supply. Concomitantly, river herring populations have nose-dived. Some can't help but question whether the successful management of the striped bass fishery has contributed to the river herring decline. With the decrease in menhaden, there is the possibility that striped bass may be relying on river herring to fill a dietary gap.

Grey Seal

Historically abundant along eastern and western shores of the Northern Atlantic Ocean, grey seals (*Halichoerus grypus*), also known as "horseheads," became a rarity by the 1960s. Their disappearance on the U.S. East Coast may have been partially due to a decline of fish that constitute their diet, but some of the downward direction of the populations was linked to conflicts with fishermen. The grey seal diet consists of commercially important groundfish such as cod and flounder, as well as salmon, skates, sand eels, and Atlantic herring. They may also munch on river herring and other anadromous fishes as the fishes move inshore before their migration to spawning areas. Seals are also a nuisance to fishermen because they sometimes get trapped in fishing gear and, as a result, fishermen are not always able to salvage the gear. Lobstermen report that the seals destroy their traps as they attempt to get at either the bait or the lobsters. One of the major industries that suffers financial losses due to seals is salmon farming; seals are attracted to concentrated source of food on the farms and are sometimes responsible for loss of salmon or damage to gear. Until 1962, because of conflicts with the fishing industry, there was a bounty on grey seals in Maine and Massachusetts. After the passage of the Marine Mammal Protection Act in 1972, grey seal populations have exploded in the Northeast as well as Maritime Canada (where the First Nation and Inuits traditionally hunted them for food and clothing). Periodically, Canada opens up a limited amount of seal hunting to cull some of the larger herds.

Grey seals can grow to large size, males up to 300 kilograms (about 600 pounds), females, up to 150 kilograms (300 pounds). They forage for meals several hours each day, with adults packing away an average of 5 kilograms (11 pounds) of fish on a daily basis. They use their razor-sharp teeth to latch onto slippery prey, which they swallow whole. The herds are now so large that seal watching has become a lucrative ecotourism industry in areas where the seals are likely to haul out of the water and can be easily observed. The recent dramatic increase in grey seals along the coasts of Maine and Massachusetts, and their presence in and near rivers that have river herring runs have prompted speculation that they may be partially responsible for river herring declines. Even though grey seals have been observed feasting on river herring, for example, at the base of the Damariscotta Fishway, it is not likely that seals alone can be responsible for river herring declines throughout their range. In a preliminary study of grey seal diet along the North Atlantic Coast (Ampela 2009), prey species were identified by examining seal scat found in haul-out areas, as well as the stomach contents of grey seals that were encountered as bycatch in fishery operations, for skeletal bones and otoliths. For the most part, adult seals preferred the sand lance, also known as sand eel (although it is not an eel), a slender fish found in large schools along the coast. Winter flounder, hake, and cod rounded out the diet. In a sample size of forty-five young seals, some fatty acid profiles of blubber samples matched those of river herring and smooth skate, but overall, a large diversity of prey species was consumed. These limited findings suggest that river herring may be consumed by young seals, but the adults have other preferences. Perhaps the seals are opportunistic and prey on river herring in spring when they are close to shore but switch their diets after the exodus of the fish into deeper marine environments.

Feed the Birds

River herring are prey for avian piscivores (fish-eating birds), including gulls and ospreys. But the most maligned bird of prey on river herring is the cormorant. The double-crested cormorant (*Phalacrocorax auritas*) is very common throughout North America and is exquisitely designed to prey on fish. It is often found, swimming and diving, along the coast and on inland waterways such as rivers, streams, lakes, and ponds, all locations where river herring can be found. They may sometimes be seen standing on shore or on a rock, with wings spread in characteristic pose for long periods of time, air-drying their feathers after a dive.

The populations of double-crested cormorants declined in the mid-1900s, attributed to the widespread use of DDT (dichlorodiphenyltrichlorethane) and other synthetic pesticides, which were banned in the early 1970s. Today, the cormorant populations are large once again, and these big black birds are sometimes considered a nuisance. This bird has made enemies of fishermen, who sometimes dub it "Black Death." Cormorants stirred up a controversy on Martha's Vineyard in 2003, when an Aquinnah Wampanoag shot 11 of the birds, which were feasting on river herring on tribal property. The tribe manages the fishery and leases the run as a source of bait. The cormorants were clearly decreasing the profitability of the lease, but in most situations it would be illegal to shoot them. Because the incident occurred on tribal land, local, state and federal laws do not apply, and thus the hands of the Environmental Police were tied. Even though the need to somehow control the cormorant population was recognized, the shooting was not condoned by the tribe's Natural Resource Officer.

In addition to river herring, cormorants feed on a variety of prey, including eels, winter flounder, trout, and even amphibians. Cormorants are protected in the United States, and the U.S. Fish and Wildlife Service is charged with their management. Officials in both the United States and Canada are considering a number of proposals to protect the birds while also protecting the fish and the fisheries, but it's difficult to manage an ecological balance. Some investigators have even predicted that as anadromous fishes such as river herring decline further, there will be a trickle-down effect on shorebirds such as cormorants; they too will be detrimentally affected. Using the cormorant as an avian piscivore model, Jones et al. (2010) examined the contribution of river herring to the diet of shorebirds. By examining the nutrient content of cormorant eggs near a herring run in coastal Connecticut, it was found that an average of 35 percent of the nutrients were derived from river herring. In eggs from cormorant colonies located further away from herring runs, the contribution of river herring to egg nutrients was lower (11 percent) but still significant. Therefore, river herring decline may be a serious problem for shorebirds, and continued declines will likely lead to far-reaching ecological consequences.

Poaching

The extent of illegal harvest of river herring is not known, but rumors abound regarding the striped bass fishermen who sneak down to the local run and, within minutes, scoop out a bucket of river herring to use as bait. Perhaps the poacher is a bait shop operator who can make a few bucks selling the herring to fishermen. No one will be the wiser unless the Environmental Police happen to be in the area. Although poaching may not be a major driver in the overall depleted status of river herring stocks, it is yet another of many factors which can contribute to the failure of those stocks to recover.

Bycatch of the Day

The United States fishing industry is fraught with controversy. Protecting their livelihood, fishermen might be described as a very vocal and opinionated group. Environmentalists, concerned about declining fish stocks, are also weighing in on the fisheries, and regulators are caught in between. Although the evidence is not complete, the impact of bycatch, i.e., the accidental taking of one species while targeting another, on the status of the river herring must be considered as one of a number of possible factors in the overall decline of local populations. The issue of river herring bycatch in the directed fisheries for Atlantic herring and Atlantic mackerel is a hot-button issue that is further explored in Chapter 9, Perils at Sea.

The Elephant in the Room

While it might be obvious that there are numerous isolated examples of human activities and industrial waste products that can alter watersheds, there is a much larger driver of ecosystem alteration that can impact numerous species, including river herring: global climate change. Some of the impacts that may have untoward consequences can be anticipated, such as change in geographic distribution of river herring as well as their predators or their prey, shifts in timing of spawning migrations, increases in the presence of nonnative species that may compete with river herring for food or habitat, more overlap in the spawning migrations of alewife and blueback herring, thus allowing the two species to interbreed (anecdotal reports of alewife/blueback herring hybrids have been circulating among fisheries biologists), potential dry down of headwater ponds and small streams, and numerous other scenarios. From an ecological perspective, there may be significant system-wide changes in watershed community structure. Of particular and immediate concern are changing seasonal patterns of rainfall and temperature. Some of these impacts may already be having an effect on the declining river herring stocks. For example, at the beginning of the twenty-first century, there have been years of heavy spring rains, which may wash away river herring gametes or eggs and precipitate the loss of certain year classes. On the other hand, there have also been some very dry summers, which can lead to conditions that prevent the out-migration of juveniles. Between the scope of human activities and the vagaries of Mother Nature, river herring face sufficient and significant challenges, contributing to their federal designation as a "species of special concern."

CHAPTER 9
Perils at Sea

Anadromous fishes such as river herring and shad present a unique challenge in terms of conservation protection. These species spend most of their year in the Atlantic Ocean and are found in offshore areas, including federal waters (the Exclusive Economic Zone or EEZ), where commercial fishing boats are on the hunt and where they may be inadvertently harvested as bycatch by directed fisheries for Atlantic herring and other species. At certain times of the year, river herring are found closer to the coast and, during spawning runs, in brackish and fresh waters, which are under state regulation. Protecting anadromous fishes in only one of their habitats may not be a sufficient conservation initiative if they are not being protected in the other habitats that they utilize. To protect a run, state regulators of the inland fisheries may be making efforts to limit, and even prohibit the catch while towns, cities, and states may be building fish ladders, taking down dams, and cleaning up rivers and streams to assure fish passage. But all these efforts may be to no avail if an entire run can be decimated by a single tow of a mid-water trawler. If the decline of river herring were solely due to a history of long exploitation and overharvest by inland fisheries, it could be argued that there should have been some rebound in the northeast stocks now that most of the inland fisheries have been closed for many years.

Several of the problems with overfishing inland, lack of passage, and pollution have been addressed. Thus, some observers believe that the dramatic and continuing decline in the return of river herring to spawning grounds indicates that something is still amiss. It is no wonder that folks are pointing their fingers to the possibility of "bycatch" as a factor, while others argue that it may be the main factor in river herring decline.

The Lawsuit on Behalf of River Herring

On April 1, 2011, a lawsuit was filed in Federal Court in Washington, D.C., against the National Marine Fisheries Service (NMFS) and the Atlantic States Marine Fisheries Commission (ASMFC), by the public interest law firm Earthjustice on behalf of fishing advocate Michael S. Flaherty of Wareham, Massachusetts, charter boat Captain Alan A. Hastbacka from Chatham, the Martha's Vineyard/Dukes County Fisherman's Association, and the environmental group Ocean River Institute. The lawsuit was an attempt to alert the federal agencies that an improved plan may be required to protect river herring, as well as the related anadromous shad, and that efforts must be stepped up to conserve and manage these anadromous species in federal waters. There is considerable fear that the industrial-scale, commercial mid-water trawl fishery for Atlantic herring and mackerel is decimating river herring and shad stocks. The lawsuit was a legal challenge to the manner in which the Atlantic herring fishery is being overseen and regulated by the federal government.

Why a Lawsuit?

The Earthjustice lawsuit contends that agencies which have been entrusted with the conservation and management of river herring and shad have failed to uphold the 1976 Magnuson-Stevens Fishery and Conservation Management Act, known more commonly as the Magnuson-Stevens Act, which calls for the protection of fishery resources in U.S. waters in the Exclusive Economic Zone which extends from three miles to two hundred miles from the U.S. coastline. The premise behind the Magnuson-Stevens Act is that, with proper management, fish stocks can be sustainably harvested. In addition to commercially important fish species, the Magnuson-Stevens Act, Public Law 94-265, also addressed anadromous species in the following section:

> The Congress finds and declares the following: (1) The fish off the coasts of the United States, the highly migratory species of the high seas, the species which dwell on or in the Continental Shelf appertaining to the United States, and the anadromous species which spawn in United States rivers or estuaries, constitute valuable and renewable natural resources. These fishery resources contribute to the food supply, economy, and health of the Nation and provide recreational opportunities.

The regulatory agencies are expected to design measures to prevent overfishing, rebuild stocks, establish annual catch limits, and, most relevant to river herring today, minimize bycatch. The lawsuit is a challenge to Amendment 4 of the New England Fishery Management Council (NEFMC) Atlantic Herring Plan, which was originally conceived with a section to monitor and minimize river herring bycatch. However, when Amendment 4 was approved in 2010, provisions to protect river herring were not included. The sections dealing with monitoring river herring bycatch and setting limits or caps for the amount of bycatch were omitted from the final version of the amendment. Instead, the amendment established catch limits for Atlantic herring with the intention of mitigating against overfishing of Atlantic herring . . . without addressing the bycatch of their anadromous relatives, the river herring. This bycatch by the Atlantic herring fleet was known to occur and had been documented by monitors.

New Rules and Regulations

On March 9, 2012, a 74-page summary judgment on the Earthjustice lawsuit against the NMFS was filed by U.S. District Judge Gladys Kesseler. The judge ruled in favor of the plaintiffs and found that the NMFS failed to take the required steps to address river herring bycatch in Amendment 4 to the Atlantic Herring Fishery Management Plan. There were three essential aspects to the ruling: Fishery Management Plans must protect all stocks that require conservation and management; these decisions should not be unreasonably delayed; and Amendment 4 failed to minimize bycatch. In her decision, the judge did not propose an immediate remedy, given the fact that a new amendment, Amendment 5, was already in the works. It would now be up to Amendment 5 to redress the omissions of Amendment 4 regarding river herring (and shad) bycatch.

The development of Amendment 5 paralleled the implementation of another amendment to a different fishery management plan. As Atlantic herring became over-fished, fishermen turned to other species. By the late 1990s, the Atlantic mackerel, *Scomber scombra,* fishery expanded. Here too, river herring are subject to the fate of bycatch. By 2012, the Mid-Atlantic Fishery Management Council (MAFMC) added Amendment 14 to the Squid, Mackerel, and Butterfish Fishery Management Plan to address river herring and shad bycatch. The amendment calls for scientific monitoring of bycatch and development of methodologies to assess and minimize bycatch, i.e., "cap" the bycatch. The goal of the 2011 Earthjustice lawsuit was the development of a similar proposal for the Atlantic Herring Fishery. When fisheries managers began working on Amendment 5 to the Atlantic Herring Management Plan, their intention was to incorporate similar accountability standards: monitoring, assessment, and action plans to cap river herring bycatch. Regarding bycatch, river herring are not the only affected species. Atlantic herring trawlers are allowed hundreds of pounds of haddock as bycatch; but, at least in the specific case of haddock, there are rules in place to limit the amount.

Bycatch 22

Can bycatch explain the recent population and stock trends reported for river herring throughout their range, such as reduced biomass, decrease in overall size of mature fish, decrease in age of maturity, and general truncation of the age and size structure of spawning fish? This is a complicated and controversial question that has pitted environmentalists against fisheries managers while forging unique and distinct bonds between some environmental organizations and recreational and commercial fishermen's associations. This question has also raised another issue: who is ultimately responsible for protecting river herring?

The management of fisheries which impact river herring is under the purview of several agencies whose responsibilities may be confusing to the general public. It certainly was confusing to me as I sought information about regulation of local herring runs and the oversight of bycatch in directed oceanic fisheries. Each town or city has an agency such as the Department of Natural Resources, which oversees the local herring run. Each state has a department or division that establishes and enforces state-wide regulations of fisheries within state waters. As part of their life history, mature river herring don't remain in their run, or in state waters, but journey out to sea after they have spawned; the young of the year also head to sea. Furthermore, the fish do not have the habit of adhering to state boundaries, so Earthjustice attorney Roger Fleming likens it to crossing into a "black hole" when they journey farther than three miles from the coast. It was in this "black hole" where foreign fishing fleets decimated river herring and other fish resources in the 1960s and it is here, as river herring enter federal waters, where the conservation and protection measures developed by each state are no longer relevant.

Troubled Waters

For management of fishes in federal waters, the National Marine Fisheries Service steps in. NMFS is a division of the federal National Oceanic and Atmospheric Administration. The East Coast arm of the NMFS is the Atlantic States Marine Fisheries Commission, which consists of representatives from 15 states. These agencies work with the U.S. Fisheries and Wildlife Service, an agency responsible for freshwater issues that can impact river herring stocks: fish passage, habitat, and water quality. In December 1993, ASMFC was given the mandate by the Atlantic Coastal Fisheries Cooperative Management Act to foster coast-wide cooperation between states to coordinate the management of migratory fishes along the Atlantic seaboard. The species involved are the following: American eel, American lobster, American shad, river herring, Atlantic croaker, Atlantic herring, Atlantic menhaden, Atlantic sturgeon, bluefish, northern shrimp, red drum, scup, black sea bass, Spanish mackerel, spotted seatrout, striped bass, summer flounder, tautog, weakfish, winter flounder, coastal sharks, horseshoe crabs, and spiny dogfish. ASMFC is further divided into the Mid-Atlantic Fisheries Management Council and the New England Fisheries Management Council.

Although there is currently no directed fishery for river herring in federal waters (EEZ), there are commercial fisheries, in particular, the Atlantic herring fishery, which, by virtue of scale and methodology, may be responsible for inadvertently contributing to the decline of river herring. In these fisheries, river herring, as well as young shad, are considered as bycatch or incidental catch. According to U.S. Code, 16 USCS § 1802 (2), Title 16. Conservation; Chapter 38. Fishery Conservation and Management 2010, the term *bycatch* is defined as "fish which are harvested in a fishery, but which are not sold or kept for personal use, and includes economic discards and regulatory discards. Such term does not include fish released alive under a recreational catch and release fishery management program." The more common understanding is that bycatch is considered to be the unintended capture of non-target species, and, by law, such bycatch must be thrown overboard, whether alive or dead. River herring are also taken as "incidental catch," which includes those that get discarded at sea (whether alive or dead) as well as those which make it into shore and get sold along with the intended harvest.

Sea Change

The pressure at sea and the bycatch problem were not always of the current magnitude. The bycatch issue first reared its head during the period when many commercially important fish species were closer to shore, and river herring were a significant bycatch for seiners who were after other fishes. However, bycatch of river herring resurfaced as a major problem when the demand for lobster bait skyrocketed in the mid 1990s, a period in which Atlantic herring was targeted for this purpose. Trawlers began to pursue the large schools of Atlantic herring off the New England coast, focusing on Georges Bank, the backside of Cape Cod, the Gulf of Maine, and the area south of Nantucket shoals. River herring sometimes mingle with or can be found near the Atlantic herring schools and can inadvertently be harvested with the intended catch. In the Mid-Atlantic, the same problem occurs with large trawlers, which target butterfish, squid, and mackerel.

A trawler may be a single vessel; some are up to 165 feet long and can hold more than a million pounds of fish. But it is much more common to see paired mid-water trawlers. These work as two vessels with a single, large net, the size of a football field, which can surround a vast school of Atlantic herring, which are subsequently raised in the nets and vacuumed into the ship's hold with little opportunity for sorting or separating fish species. Another fishing technique that has the potential to catch river herring is purse seining, which also does not discriminate between species of fish. Purse seines are large floating nets, equipped with weights on the bottom. Also on the bottom are loops through which line or wire is threaded. When a school of fish is spotted, the ship encircles the school with the net and then the line is pulled to close the net at the bottom. This cinching method was used to close women's purses (hence the name purse seine) before the days of the zipper. Bottom trawling is another fishing method that has the potential to accidentally capture river herring. Whether a trawler net or a purse seine, the entire net is hauled aboard, where, mingled among the targeted catch, river herring are sometimes also trapped.

Although this might not be a common event, a single haul of a large midwater trawler or of paired trawlers may contain more river herring than typically migrate to a particular spawning ground. Thus, for a specific river or stream, an entire annual run may be lost. Some maintain that river herring bycatch would be a non-issue if it weren't for the scale of the Atlantic herring fishery. When Atlantic herring were fished for food, the demand was not comparable to today's clamor for lobster, and thus lobster bait. This may also be an ominous sign for the future of Atlantic herring, whose stocks also are declining.

Assessing the Extent of Bycatch: The Sampling Dilemma

There are a number of opportunities and methods to assess the magnitude of river herring bycatch in the directed Atlantic fisheries. Granted, this is a difficult, almost unimaginable, assignment for a variety of reasons, logistical as well as statistical. The overall system is not without flaws. These methods produce statistics that are categorized as fisheries-dependent data (harvest and bycatch records from fishing log books collected by the fishermen during their trip) and fisheries-independent data. There are two occasions in the fishing process where monitoring can occur to deliver fisheries-independent data: at sea and at portside. The overall system has several inherent problems; voluntary information, such as that obtained from fishing log books, most likely underreport river herring bycatch, especially if the catch is not sampled or scrutinized, while the at-sea observer program currently has low rates of coverage on vessels. Port-side observers get to see only the bycatch that is not discarded/dumped at sea.

The NMFS uses an at-sea (on board) observer program in many of the fisheries, but it is virtually impossible to survey the entire hold of a large-volume trawler at sea. The Northeast Atlantic herring fleet consists of about a hundred vessels, which remain at sea for several days at a time. About a dozen of the vessels are responsible for the bulk of the catch (over 90 percent). Given resources available for sampling bycatch, it has been impossible for every ship to have trained fisheries observers on board. Furthermore, when a ship encounters a school, entire nets-full of fish are pumped into the hold, with no time or opportunity to identify, count, or weigh individual fish; thus batch sampling is the norm. Batch sampling involves pulling off a small subset of the catch and going through it systematically, fish by fish. Each fish is identified, and other data are collected as needed for documentation and analysis. The variable issues inherent in at-sea sampling are that not all the vessels are covered by observers 100 percent of the time or in 100 percent of the fishing areas. A net may contain bycatch on one day, but not the next; it may contain bycatch when the vessel fishes in one area, but not another. An additional practice of the Atlantic herring fleet that is problematic regarding bycatch is that, for various reasons, the vessels sometimes do not bring all catch on board, and may "dump" contents of their nets, which may contain dead fish and/or bycatch, before observers have an opportunity to sample it.

The other method of estimating bycatch, dockside or portside sampling, is conducted by state Divisions of Marine Fisheries or Departments of Natural Resources. This method assesses the bycatch that is landed as it is being offloaded into food- or bait-processing facilities. I did not have the opportunity to observe this process, but according to Cieri et al. (2008), "Samplers position themselves at the point of entry into the facility along an assembly line or at the base of the hoppers where fish are unloaded. Sampling is conducted before grading or sorting of the catch occurs. All bycatch is removed from the assembly line or hopper and is placed in bushel baskets or buckets specific to each species." A protocol is followed so that length and weight of each fish is recorded. The identification of the vessel is confidential; only the vessel gear type, time of year, and port of landing are entered into the database. Although this method can give an estimate of river herring bycatch, a limited number of lots of fish are sampled and thus extrapolation to the entire fleet may not be accurate.

Schooling for the School

I was very impressed when I visited NOAA's Northeast Fisheries Observer Program (FOP) facility in East Falmouth, on Cape Cod. This facility is the training ground for fisheries observers, i.e., individuals who accompany fishing vessels from Maine to North Carolina to collect data for the National Marine Fisheries Service. The observers are recruited by A.I.S. (Accuracy, Integrity, Service), Inc. which contracts FOP to provide three-week intensive training sessions in several monitoring programs, one of which serves the Atlantic herring fleet. The bulk of the observers awaiting certification tend to be young, college-educated individuals with degrees in biology who don't mind unpredictable schedules, rolling seas, and unglamorous work. Those who accompany herring fishing vessels undergo advanced training for the high-volume nature of the fishery before they can be certified. The observers spend one to four days at a time at sea, performing vessel safety checks, collecting samples of fishes that are vacuumed into the holds of the trawlers, sorting through the samples, and collecting data about the types and sizes of fishes in the samples and the type of gear being used. I watched the action unfold on a large flat-screen monitor. Amy Van Atten, Branch Chief of the FOP, showed me a video of some paired trawlers at work, so that I could appreciate the volume and speed at which fish are being pumped on board. Using electronic fish finders, the vessels position themselves to maximize the catch. After the fish are collected in the massive net, they are loaded aboard one of the vessels. To sample the silver stream of Atlantic herring being pumped on board, an observer positions buckets at periodic time intervals to catch samples which would later be sorted. The frenetic pace of the operation leaves little time for rest or contemplation; the observers must be very attentive and aware of their position and actions, lest they be sucked into the hold and buried by tons of fish.

I wondered how the observers were trained to distinguish between Atlantic herring, alewives, and blueback herring. Van Atten took me into the large warehouse section of the FOP facility where fishes were arriving for I.D. verification. To make sure the observers are correctly identifying fish species, the on-board observers are required to submit small subsamples to the FOP for confirmation. The rubric for river herring I.D. was explained: first, the observers look for the black shoulder spot and a dark iridescent blue to silver back, to distinguish river herring from Atlantic herring. Another telling criterion is the sawtooth belly on river herring, easily observed by touch, running a gloved finger along the ventral surface from the anal fin in a forward direction. Then, to distinguish between alewives and bluebacks, the fish are cracked open in the middle so that the color of their gut cavity lining can be noted. The alewife I observed had a pale pinkish gut lining, while that of the blueback on the table was completely black. I am happy to report that the at-sea observer made the correct call and the staff at the FOP was able to verify the identification of the fish. The at-sea observers provide valuable data which are then used to assess the amount of bycatch and the effect of different gear types in different locations and during different time frames. They provide a valuable service, but, they are not required to accompany every vessel in the herring fishery.

Eyes on the Prize

Mike Armstrong, of the Massachusetts Division of Marine Fisheries, estimates that, during 2010, the Atlantic herring fleet, 85 percent of which lands in New Bedford, Fall River, and Gloucester, collected 1.5 million river herring (about 700,000 pounds) as bycatch. Using data from Maine and Massachusetts, where bycatch was estimated using two techniques: at-sea subsampling of catches (a set number of baskets per haul) and portside sampling of entire lots trucked into processing plants and bait dealerships (which may be a subset of fish from a particular vessel or vessels), Cieri et al. (2008) estimated that the river herring bycatch in the Atlantic herring fishery ranged from 0.2–1.7 million pounds annually between 2004 and 2007. In one year, the bycatch was highest in bottom trawls; in other years, single and pair trawls were the worst offenders, while purse seines had the least river herring bycatch. Overall, the study showed considerable variability from year to year and gear type to gear type, but suggested that even though river herring bycatch may be low in terms of percent of total catch, it has the potential to be "substantial" due to the high volume of the Atlantic herring fishery.

In 2009, Becker submitted a report to the Atlantic Cooperative Statistics Program using lot sampling of a selected number of entire Atlantic herring catches brought to processing plants from individual fishing vessels from July 1, 2007, through October 31, 2008. This study, which compared four types of gear, captured 5 percent of weekly Atlantic herring landings and 2 percent of

Atlantic mackerel landings in up to twelve sites from Maine to southern New Jersey. In reporting about the Atlantic herring fishery, it was revealed that 1.5 percent of the haul delivered to the processing plants was bycatch, including twenty-five different species. Of the bycatch, most of the species were fishes, although a few crabs and lobsters were also discovered. An important finding was that 78 percent of the bycatch was Atlantic mackerel while about 7.9 percent was river herring. In addition, pair trawls accounted for the largest chunk of the bycatch: 45 percent in 2006–2007 and 50.9 percent in 2007–2008, followed by purse seines (30–40 percent), and single trawls (18–25 percent). About 1 percent of the bycatch was attributed to fixed gear. Becker (2009) admonishes, "It is inappropriate to extrapolate this data to the entire herring and mackerel fishery." But once again, it is important to emphasize that, given the high-volume nature of the fishery, even a small percentage of river herring bycatch can translate into a large number of fish.

Bycatch Monitoring and Marine Conservation

Many observers believe that in order for conservation initiatives to be put into place, and more important, in order for them to work, observation and documentation of bycatch must occur and it must occur regularly. Limited, selective sampling and subsequent extrapolation of bycatch data can be quite misleading. River herring do not stay in one location; they move along the coast, being present in one area during a particular time period and then moving on to another area. It is known that, depending on the time of year, river herring bycatch is significant in some locations, but not in others. Scientists have attempted to track the temporal and geographical movements of river herring using a variety of methodologies. For example, analyzing bycatch data from direct on-board observation and vessel trip reports and conducting trawling surveys have made it possible to deduce that there may be specific patterns to river herring movements at sea (Cournane and Correia 2010a). These patterns can be statistically assessed, and if the patterns are documented, significant, and predictable, it may be possible to manage the Atlantic herring fleet so that trawlers avoid these seasonal "hot spots." This can be done by implementing temporal, gear, or seasonal closures in certain locations. Other types of management may involve close monitoring of bycatch at sea, so that if river herring are turning up in the catch, the ships may be required to move along to another location, a so-called "real-time" avoidance technique (Cournane and Correia 2010b). A similar bycatch avoidance system method has been adapted for a pilot program developed by MADMF and the School for Marine Science and Technology (SMAST) at UMass Dartmouth.

As the process for drafting and discussion of Amendments 4 and 5 to the Atlantic Herring Fishery Management Plan continued, there was a great deal of public comment. On September 7, 2011, a coalition of watershed organizations, environmental groups, and concerned citizens sent a request to the governor of Massachusetts that he direct the Massachusetts representatives on the ASMFC to support 100 percent at-sea monitoring on industrial midwater trawlers, protect river herring by imposing area and time closures, and require that all catch be brought on board and accurately sampled by independent observers.

On November 8, 2011, the Herring Alliance, on behalf of its forty-two member organizations, petitioned the NEFMC to impose a catch cap for river herring and include it as part of Amendment 5. A cap is another type of management initiative, which would control the amount of allowable alewife and blueback herring bycatch. Inherent to the problem of setting a cap limit is having sufficient scientific data to know what the cap should be. For these purposes, scientists typically use a tool known as a stock assessment. For river herring, there are insufficient data to provide information that could be used in statistical modeling studies to establish a reliable stock assessment. Should the cap data be based on counts observed at individual runs, which reflect specific subpopulations? Or, should the data be based on coast-wide assessment, and if so, how would that be accomplished? Therefore, the major problem in setting bycatch limits is lack of information about population and subpopulation (metapopulation) size and structure to determine what level of bycatch could be permissible without contributing to lack of sustainability of river herring populations. Scientific investigations are currently under way to shed some light on these issues. Using molecular methods, researchers are attempting to determine genetic profiles of river herring collected from individual runs while other laboratories are focusing on otolith microstructure and microchemistry as techniques to identify separate stocks. Once there are criteria to assign each fish to a specific stock, it will be possible to determine the impact of bycatch on individual runs.

Subsections of the proposed Amendment 5 demonstrated continuing progress toward minimizing river herring bycatch. For example, industrial trawlers would be required to carry and fund independent observers on 100 percent of fishing trips. To allow a more correct assessment of bycatch, there was a requirement that the weight of the catch be determined, rather than estimated. The trawlers would be allowed slippage, i.e., dumping the contents of the net before the observers can sample the catch, only ten times. If they exceed this maximum, vessels would be required to return to port. Furthermore, as the river herring stock assessment continues, river herring catch caps would be specified in future adjustments to the plan. The NMFS asked for comments on the proposed rule to implement Amendment 5 by July 18, 2013. With many years of planning and work by the NEFMC and support by many agencies, nonprofits, and interested individuals, everyone was optimistic that the proposed conservation measures would soon be in place.

It was a grave disappointment to river herring advocates and many commercial fishermen when, on July 19, 2013, the announcement was made by the NMFS that many of the recommendations in Amendment 5 to limit river herring bycatch would not be approved. The NMFS essentially discarded most of the "accountability" provisions in Amendment 5 that would require additional work and expenditure by the Atlantic herring fleet: measures that would result in protecting river herring. The NMFS claimed that there were legal concerns about implementing the measures. Because the NEFMC did not specify how the observers were to be funded, the NFMS did not believe that the proposal was sufficiently developed. Although some measures in Amendment 5 will be adopted to manage the Atlantic herring fishery, there has essentially been a five-year setback in terms of addressing river herring bycatch. It will only be a matter of time before there will be a new lawsuit and a disturbing repetition of the entire process of amending a fishery management plan.

CHAPTER 10
River Reruns

In Plymouth, Massachusetts, where some of the first mill dams in the United States were built, alewife runs continued to decline even after fishways were constructed. Over the years, some of the fishways entered a state of disrepair, became virtually nonfunctional, and constituted a roadblock for any adult river herring with a mindset to reproduce. So, by the early twentieth century, a strategy was adopted to boost the alewife population in the region where they were first harvested by Native Indians and later by colonists: fish were caught near the mill sites and carted to spawning grounds in the Billington Sea. The practice of transporting fish to detour around dams, so-called trap and truck operations, continued along Town Brook in Plymouth until new fishways were constructed, and this hands-on technique is still being used in select locations in the Northeast to restock or supplement historic herring runs.

Trapping and trucking has its merits as a last resort, but is not a very efficient or reliable method for restoring river herring populations. Several other approaches have been employed, some of which are feasible only at specific sites or for specific runs. Assuming that water quality problems such as pollution have been mitigated, there are two general approaches that are used to restore river herring stocks in the inland portion of their life cycle: bypassing dams or removing them altogether. Each initiative has its own challenges, but even if the effort is not completely successful in restoring runs to their historic levels, at least a cadre of upstream migrants will be assured an unimpeded route to spawning grounds.

Lift Days

Not far from Boston, Mystic Lakes are man-made bodies of water that originally were part of the upper reaches of the Mystic River. The lakes were formed when dams were constructed on the river to serve leather mills and to provide water to neighboring towns. Although river herring that spawn throughout Massachusetts are primarily alewives (perhaps 90–95 percent), the Mystic River system stands out as a run for blueback herring. With the construction of a dam and the creation of Upper and Lower Mystic Lakes the bluebacks lost access to spawning habitat in the Upper Lake. In 2005, "Give Fish a Lift" became the battle cry for the newly formed river herring bucket brigade. From 2005 to 2009, volunteers worked tirelessly, standing on rocks and islands of concrete at the base of the dam to deftly scoop fish from below the dam, place them in buckets, and manually hoist them above, where they entered a chute that delivered them to the Upper Lake. The manual lift provided adults with a ticket to spawning grounds, but problems would some times arise after dry summers when there would not be enough water for the juveniles to get over the dam. The bucket brigade was not a workable solution for the young of the year, so the juveniles were at the mercy of the weather, able to emigrate only if there was significant rainfall.

For a variety of reasons, the dam could not be removed, but there was potential for 165 acres of spawning habitat in the Upper Lake. After much planning, a new dam was completed in spring 2011. With this new incarnation of the dam, an appropriate fish ladder was part of the project, along with the inclusion of structures to allow eels to pass over the dam. The juveniles no longer have to wait for heavy rain; the new dam also provides a structure for juvenile downstream passage. The dam / fish ladder at Mystic Lakes is one of many projects that are completed, in the works, or in the planning stages, with the goal of restoring some of the Northeast's historic herring runs, many of which function sub-optimally, if at all.

Similar to the fish lift effort on the Mystic River, River Herring Rescue is a small, grassroots event organized by New Jersey fishermen who are counting on the health and abundance of game fish such as striped bass to support shorefront businesses such as boat dealerships, fishing charters, and bait and tackle shops. The fishermen are keenly aware that the abundance and condition of striped bass is linked to their forage base. On "Lift Day," volunteers net river herring at the base of dams or other impediments and lift them over the top to help them on their way to spawning grounds. It is probably very tempting to keep a few for bait, so the event is closely monitored by fish, game, and wildlife bureaus to ensure that no poaching occurs.

Lifting fish is a labor-intensive strategy but sometimes it's the only reasonable option to maintain a run until more permanent measures are in place. For several years, a trap and truck operation has kept the run in Pembroke, Massachusetts, alive after a restocking program brought fish back to the North River. Fish receive an assist on their twelve-mile odyssey by getting a one-way ticket to their spawning area in Oldham Lake. The Glover Mill Dam, slated for repairs, would be a dead end for fish if volunteers did not herd them to the base of the dam, scoop them into nets, and load them into a large tank at the back of a truck. A five-mile ride brings the migrants to the lake.

In some cases, fish lift operations are accompanied by a restocking program, as is the case at the Main Street Dam in Wakefield, Rhode Island. Here, some of the fish that get a lift are trucked by the Rhode Island Department of Environmental Management to other rivers or streams to kick-start or supplement runs. Over the past decade, many of Maine's rivers have been stocked by so-called trap and truck or truck and transfer operations. The Maine Department of Natural Resources has used these strategies, for example, on the Androscoggin, Union, and Sebasticook Rivers, to detour river herring around impediments such as hydroelectric dams. In the case of high dams, such as those on the Connecticut and Merrimac Rivers, hydraulic, rather than manual, power is needed to assist the fish. For example, the fish lift at the Holyoke Dam elevates fish 52 feet (18.85 meters), which is out of the range of manual intervention. Examples of hydraulic fish lifts in Maine include the elevators at the Lockwood Dam, constructed in 2006 to allow diadromous fishes access to the upper Kennebec River, and the Skelton Dam fish lift (the largest in Maine) on the Saco River in Dayton, constructed in 2001 at a cost of $6 million by Florida Power and Light Energy. Some of the river herring that ride the Skelton Dam elevator are destined for restocking programs which are overseen by state officials.

"Rights" of Passage

Lifting, trapping, and trucking are not always feasible interventions. These efforts require close seasonal monitoring, scheduling, and a volunteer effort, or a major investment in a fish elevator. Although these relocation efforts are helpful, they are not without negative side effects: because their homing instinct is interrupted, fish may become disoriented, delay their spawning, and even succumb to disease or otherwise suffer mortality. As a consequence, most efforts to allow passage rely on the fish to get themselves from point A to point B. However, it is important to bear in mind that "one size fits all," does not apply to constructed fishways. As early as the 1920s, Belding wrote about fishways, "The reason why any single type is not uniformly successful is that in each case it must be adapted to the locality where installed, since different situations demand different kinds of fishways to meet their requirements" (Belding 1920, 59). He further noted, "A successful fishway which will take all species of anadromous fish has never been invented" (59). Such is the problem in trying to design a structure to allow fishes to bypass a dam or other obstruction. Many of the older fish ladders were actually designed for salmon. It is only recently that attention has been focused on the particular structural requirements suitable for alewives and blueback herring. Belding outlined the requirements for fishway design that would appease dam owners as well as allow passage of river herring: the fish must have easy, rapid passage (this would require uniform flow of water, gradual ascent, and absence of high barriers), there must be minimum sacrifice of water (to make dam owners agreeable), the fish must readily be directed to the entrance, and the fishway must be solidly constructed (so it won't fall apart if it is intended to be a permanent structure), or removable (so it can be taken down after spring migration).

Every location where a fishway is needed presents a unique situation in terms of the overall length, slope, height, surrounding features, species of fish, and goal of the project. Overall, two basic aesthetics can be observed in our northeast fishways: one concept calls for a natural-looking structure, appearing as though it had always been situated in a specific location. These "nature-like" fishways are made of materials such as rock, timbers, and gravel; they tend to be designed for a specific site and they attempt to pass all species of fishes endemic to the river or stream. In contrast, there are a number of

10.1. Weir and pool fishway design, Damariscotta Mills, Maine (photo: Barbara Brennessel)

fishway designs that look more industrial, taking the shape of ramps or ladders, constructed from wood, aluminum, or concrete.

Two basic engineering designs underlie the creation of fishways: weir and pool (or pool and weir), and baffle or Denil-type, each with many structural variations. The earliest types of fishway were among the weir and pool variety, constructed by stacking stones to make small dams, walling off areas of a stream at various intervals, thus creating a series of pools, each at a higher elevation along the stream than the one before it. Nature-like fishways are built on the weir and pool principle. A portion of the stone and mortar Damariscotta Mills fishway (fig. 10.1) and the Stony Brook Run in Brewster (see fig. 1.5) are modeled on the weir and pool concept. However, some of the more modern weir and pool fishways are not nature-like at all; they are constructed using a long rectangular cement chute, with cement or wooden walls perpendicular to the foundation to create pools. Sometimes, vertical slots are formed within the walls to control the water flow between each pool and to allow fish to pass from pool to pool more easily (fig. 10.2).

The second design for a fishway is a long chute, which utilizes a series of baffles or vanes at regular intervals to create turbulence and decrease the

10.2. Slotted weir/pool fishway, Middleborough, Mass. (photo: Barbara Brennessel)

velocity of water. Most are modeled on the 1909 roughened ramp design of a Belgian scientist, G. Denil, and are thus called Denil fishways. The Alaskan steeppass (figs. 10.3 and 10.4), which has been adapted for use with river herring and shad, is a Denil-type fishway in which the placement and angle of the baffles differ somewhat from Denil's original design. The steeppass is a prefabricated aluminum structure, originally designed for salmon and extensively used in remote areas of Alaska; it has also been used for river herring, particularly when there is only room for a narrow structure.

There are many important components to a successful fishway, starting with the design of the entrance. The flow velocity must be within a certain range (not too fast, not too slow) to lure the fish in the right direction. The height, length, and slope of the structure must be carefully considered. For long fish passage structures, it is critical to incorporate resting pockets, where fish can take a brief break. If the structure consists of a steplike system of pools or weirs, the height of each "step" must be carefully considered; most designs limit the height to 6 inches, although some fishways incorporate a few 12-inch drops.

Abby Franklin, Anadromous Fish Restoration Project Manager within the Cape Cod Conservation District, works with the Cape Cod Water Resources

10.3. Alaska steeppass fishway, Parker River, Byfield, Mass. (NOAA Restoration Center, Louise Kane)

Restoration Project to restore fishways on the Cape. Franklin is overseeing a plethora of projects to restore or enhance fish passage, made possible through funding from the U.S. Department of Agriculture through its Natural Resources Conservation Service. She wrote to me that the "type of water flow is important . . . and will influence how well a herring will swim from pool to pool. Herring like 'streaming' flow where the water travels from pool to pool across the surface. They do not like 'plunging' flow where the water flows over the weir, then plunges down to the bottom of the pool and then circulates back up. . . ."

When one builds a fishway, it is important to know how river herring and other fishes respond to additional factors, such as turbulence and the total number of consecutive steps or drops. Fishway designs must also take into account the potential exhaustion factor, by considering the overall distance that must be traveled from ocean to spawning grounds. Some systems have a multitude of impediments; thus each individual fishway cannot be so challenging that the fish run out of steam before they encounter the next challenge on their way to spawning habitat.

And we must not forget that most of the river herring that go up must also come down, so it is also crucial to give some thought to how the spent

10.4. Interior construction: Alaska steeppass fishway, Parker River, Byfield, Mass. (NOAA Restoration Center, Louise Kane)

adults and, later, the juveniles will get back to the ocean. Builders of older fishways gave no thought to this issue and banked on the water flow to direct downstream movement over the dam. This hit-or-miss transport of fish downstream causes problems for adults, which do not favor being swept over sharp crests, and thus can delay their return to the ocean, while also causing problems for the emigrating young, which suffer fatal crashes on rocks or structures at the base when they take headers over high dams, or can't emigrate at all if there is not enough water flow.

To learn more about the intricacies of fishway design, I traveled to western Massachusetts to visit the Silvio Conte Anadromous Fish Laboratory near Turners Falls, on the Connecticut River. The research complex, established by the U.S. Geological Survey, Department of the Interior, was named after Congressman Conte, an avid fishermen with concerns about the environment. With a scientific staff committed to the study of the ecology, physiology, and behavior of anadromous fishes, including salmon, shad, sturgeon, and river herring, the facility provides opportunities to design and test various models and structures for fish passage. Ted Castro-Santos, whose specialty areas include radiotelemetry to assess fish passage, biomechanics and behavior of migratory fishes, and the bioengineering of fish passage structures, gave me a tour.

Some of the grounds and the indoor spaces in the facility resemble a water park for fish, replete with tanks, tunnels, tubes, and slides. But these structures are not there to provide fun and exercise for fish; they are well-planned designs such as flumes and fish ladders, engineered to study the ability of fish to orient and swim under various experimental conditions. Some of the fish passage designs are being tested with water from the river flowing through the facility and diverted over various design features. One design looked like a natural stream, but it was more akin to a movie set; the rocks were fake, made of light material which could be repositioned to test the effectiveness of their placement within the "set." Variables such as slope, placement of rocks, and number and location of resting pools could be examined by placing tagged fish in the "downstream" portion of the structure and following their progress upstream with high tech detection gear. It is the mission of the Conte Laboratory to use the principles of engineering and fish behavior and ecology to design and test the best possible fishway structures that would pass all species of fishes that normally migrate to a particular river or stream. The engineering concepts that are involved in the fishway designs were way over my head, but the take-home message was clear and echoed that of Belding: each species is unique and a fishway designed for one type of fish may not work for a different type.

The Role of Dam Removal in Restoration

Construction of fishways is an important consideration in plans to restore or revitalize river herring runs where dams or other impediments are in place, but some proponents of restoration want to "go for the jugular" of the dam by taking it down. Even though well-designed fish ladders may provide passage, the structures must be permitted, maintained, repaired, and periodically replaced. Based on the trajectory of restoration projects recently completed or in progress, dam removal, when feasible, is front and center, being cited as the single most useful action that can be undertaken to promote the recovery of river habitats and accelerate the resurgence of river herring populations (Limburg and Waldman 2009).

Yet, however appealing it might be to breach or remove a dam, it may not always be possible, even if the dam no longer serves its original intended function. For example, contaminated sediments, trapped in the lake or impoundment above the dam. may flush downriver if the dam is removed. Sometimes it's not possible to remove a dam because of impacts to landowners which cannot be easily mitigated, while at other times, dam removal will compromise the ability of a region to deliver hydropower to its inhabitants. Occasionally there will be resistance to dam removal by nearby landowners who value the esthetic and recreational opportunities afforded by the lakes and ponds formed as a result of impoundment of the river. This resistance sometimes pits waterfront property owners against the communities in the rest of the watershed.

Although dam removal can be a very expensive proposition, it may be less costly in the long term for dam owners to remove old, failing, nonfunctional dams than to keep them in good condition. For example, Massachusetts laws pertaining to dam safety, revised in 2002, impose new responsibilities on dam owners that may provide incentive to take down the dams. Owners are now required to register their dams, as well as to arrange for periodic safety inspections. Owners must maintain dam structure and prepare Emergency Action Plans for dams that are classified as having "High or Significant

Hazard Potential." Therefore, dams may have very high maintenance costs and incur unwanted liability on their owners.

Which dams are possible to remove? What are the ecological consequences? How can the project be accomplished? Should the dam be completely removed or partially breached? Which of the many worthwhile projects should be scheduled first? Many engineers, restoration project managers, and scientists are attempting to answer these questions. Prioritization of potential projects requires the identification and understanding of many complex issues: there are many economic, ecologic, and logistic aspects to these projects, whether they involve a small historic structure or a large hydropower plant. Most dam removal projects take years, and even decades, with conception of the project, planning, scientific studies, and permitting taking the bulk of the up-front time. These complex projects generally require the partnership of a multitude of cooperating agencies, permitting hurdles, and major sources of funding. Many unexpected issues and controversies have surfaced as dam removal projects enter the planning stages. I will highlight a number of river and estuary restoration projects up and down the northeast coast. Many of these restoration plans are embedded in large-scale watershed protection and conservation projects in which dam removal is one of many components. Most of the plans require construction of some type of fishway in addition to the removal of the offending dam. A number of the projects illuminate a sampling of the issues and problems with attempts to restore fish passage or remove dams and other structures, large and small, which impair the ability of river herring to find suitable spawning habitat.

The Maine Event

In 1945, with the assumption that streams could be "returned to production," Rounsefell and Stringer recommended measures to restore alewife runs on seventeen streams in Maine by building or modifying fishways, restocking, and "proper management methods" (Rounsefell and Stringer 1945). This imperative has been taken on by members of the Alewife Harvesters of Maine. Many members of the group are fishermen whose livelihoods are at least partially dependent on strong, healthy river herring runs. The group prides itself as having members who are stewards of their runs. They are involved in many projects, including fish ladder construction to allow passage around dams as well as advocating for their industry. They have been working with the Maine DNR by providing samples and data that are used in analysis of population size and structure. However, in order for Maine streams to return to production, some of Maine's major watersheds require triage.

The U.S. Fisheries and Wildlife Service (USFWS) established the Gulf of Maine Coastal Program (GOMCP), one of twenty-one of such coastal programs nationwide with the mission "to build partnerships to identify, protect and restore nationally significant habitat for fish, wildlife and people" (USFWS GOMCP 2007). GOMCP has put together federal, state, and NGO groups in a number of partnerships and programs to protect and restore diadromous fish habitat in Maine. Some of their projects enlist the assistance of land trusts, watershed groups, municipalities, academic institutions, owners of timberlands, agricultural interests, such as blueberry farmers, and many other stakeholders. It is important to emphasize the "protection" function of some of the projects because protection of habitat provides the opportunity to maintain the connectivity of existing rivers and riparian habitat and prevent their degradation to the point where a rehabilitation or restoration program must be put in place. It's usually better and less expensive to prevent a problem that to devise a remedy later on.

River herring, along with shad, Atlantic salmon, Atlantic sturgeon, shortnose sturgeon, rainbow smelt, striped bass, Atlantic tomcod, sea lamprey, and American eel, will be the beneficiaries of these efforts to restore productivity in Maine's watersheds. In addition to enhancing the stocks of anadromous fishes, these restoration projects will result in millions of juvenile river herring being delivered to the Gulf of Maine each year, with the added bonus of providing important forage for groundfish stocks.

Working with landowners and identifying priority sites, the GOMCP has initiated over 135 projects since 1998. Many of these will have a positive impact on river herring, including "removing tidal restrictions and dams, repairing and renovating old fishways, installing new fishways, controlling erosion from nearby uplands (i.e. gravel pit remediation, riparian plantings/fencing), designing and installing appropriately sized and located culverts, providing temporary 'loaner bridges' to timber operators to minimize permanent road construction activities over streams, providing support to aquaculture and blueberry operators" (USFWS GOMCP 2007).

The Kennebec and Penobscot Rivers are the subjects of several of the massive projects undertaken by large partnerships in Maine. These watersheds cover great expanses within the state. The Edwards Dam was completed in 1838 on the Kennebec at the head of the tide in Maine's capital, Augusta. Over the years, the dam powered seven sawmills, a gristmill, machine shops, and, eventually, a textile mill. With the decline of the need for the dam by those industries, the dam became a hydropower source in 1884, and its proposed removal became part of a much larger plan.

The acceptance of the Lower Kennebec River Comprehensive Hydropower Settlement Accord represented a nonregulatory, voluntary, multiagency partnership, which resulted in a number of actions to restore river herring to the Kennebec. The first step involved stocking of river herring in the lakes and ponds in the Sebasticook River watershed and the Seven Mile Stream drainage, important Kennebec tributary habitats. The next step was possible when the Edwards Dam was accepted on January 1, 1999, as a "gift" from its owner, the Edwards Manufacturing Co. This resulted in the removal of the dam, which allowed some diadromous fish migration, but did not directly benefit river herring, whose historic spawning habitat was much farther upriver. To address river herring, the project therefore included other dams and multiple fishway construction projects to pave the path for river herring to access historic spawning grounds in the Sebasticook River. Several additional projects were part of the overall restoration plan. These included mitigations at five non-hydropower dams in the upper watershed and providing passage at two downstream hydropower plants at Benton Falls and Burnham. The dams at Archers Mills (a nineteenth-century relic, discovered during the course of the project) and Guilford were removed; an Alaskan steeppass was placed at the Pleasant Lake Dam; a fishway was installed at the Sebasticook Lake Dam; and two fishways were installed at the Plymouth dam. In 2008, after five years of legal battles, the Fort Halifax Dam, the last impediment for river herring, was removed. The dam was owned by Florida Power and Light Energy Hydro, LLC, which decided that the dam was not generating enough revenue to justify the installation of a fish lift. From the standpoint of the dam owner, it made economic sense to take it down. The major opponent to dam removal was a group of adjacent landowners, calling themselves "Save the Sebasticook," who wanted to maintain the impoundment created by the dam. They were living on a man-made lake, resulting from an impoundment that the dam created, and were wary of the changes that would occur after its removal.

Another partnership seeks to restore runs for river herring and other diadromous fishes in the Penobscot River. The Natural Resource Council of

Maine is a member of a group of stakeholders which include the Penobscot Indian Nation, the Atlantic Salmon Federation, the Penobscot River Restoration Trust, federal and state agencies, dam owner, PPL Corp., and many conservation organizations, which have banded together to restore 500 miles of the Penobscot River, while still maintaining power production at hydroelectric facilities. The Penobscot watershed is the largest in Maine, and it provides the greatest freshwater input into the Gulf of Maine. It is projected that the project will have a positive effect on the entire ecosystem—the river as well as its riparian corridor—benefiting all manner of wildlife as well as overall water quality while also providing increased recreational opportunities.

There is no debate about the need for hydroelectric power generation on the Penobscot; the issue is how to maximize production of hydropower while allowing passage and addressing downstream fish mortality at dam sites. The Penobscot project involves the purchase of three dams from PPL Corp. with the intention of removing the Veazie and Great Works Dams, the two dams closest to the ocean, and thus the first impediments river herring would encounter, while constructing a new natural-type river channel fish bypass at the Howland Dam. So, the large-scale project designed for the Penobscot represents a compromise, rather than a complete restoration. As with other similar projects, there are a number of opponents. In the case of the Penobscot, some argue that allowing passage of native sea run species will also present opportunities for invasive fish, such as the carnivorous northern pike (*Esox lucius*) to make their way to the upper reaches of the system and upset the existing ecological balance.

Alewife Rollercoaster

In the northeast sector of Maine, an important alewife restoration project has the potential to return significant numbers of alewife to historic habitat, but the project was on hold for decades because of a controversy that raged on the U.S.-Canadian border: native fish vs. introduced fish. There are those who see the ecological and economic importance of restoring alewife runs on the St. Croix River and those who see the alewife as a serious threat to their own ability to make a living on the river. The St. Croix, running along approximately 100 miles of the international boundary between the United States and Canada, flows through northeastern Maine and south-western New Brunswick. Beginning in the late 1700s, a series of dams were constructed on the river, but it was the construction of the Union Dam, in 1836, that marked the declining trend of alewife migration. At the same time that alewife populations began to struggle, small mouth bass (*Micropterus dolomieu*), a popular recreational freshwater sportsfish, were introduced (reviewed by Willis 2009).

After completion of initiatives to clean up the river and install fish passage structures on some of the dams, it looked as if the alewives were rebounding. But, beginning in the late 1980s and made official by 1995 Maine legislation, fishways were blocked on parts of the St. Croix, resulting in barriers to alewife migration and another serious decline in the run. Canadian officials, environmental organizations, management agencies, and marine fisheries departments were all opposed but powerless to reverse the decision.

What was the rationale for blocking passage of alewives? It turns out that those small mouth bass, introduced over a century ago, are the foundation of economic stability for a small cadre of recreational fishing guides, who blamed the alewives for declines in their target sport fish in areas of the St. Croix such as Spednic Lake. This lake is actually a very large impoundment on the St. Croix formed between the Vanceboro and Forest City dams. Without any scientific evidence, some claimed that the mature alewives were causing problems by altering habitat, affecting water quality, and eating young of the year small mouth bass. Young of the year alewives were blamed for competing with young of the year bass for food. However, concomitant with the presence of alewives and the decline in small mouth bass, there were also massive drawdowns of water in Spednic Lake by a hydroelectric company, resulting in potential loss of habitat for bass fry. With both species declining, managers were forced to take action. Which fish would be the focus of restoration? What steps should be taken?

Different management agencies and stakeholders took competing approaches to resolve the controversy, and some lawmakers, caught in the middle, were asked to take sides. For example, in 2001, conservation groups made a failed attempt to have the 1995 law repealed. Despite evidence to the contrary (Flagg 2007), the fishing guides have made claims that alewives are not native to the area on the St. Croix where they conducted their businesses, and thus should not be protected or reintroduced. When the 1995 act resulted in closure of the Woodland fishway, frustrated Canadian agencies began trapping and trucking alewives from the Milltown area to the

Woodland impoundment. Scientists gathered evidence to examine the interaction between the two fish species, sometimes with conflicting conclusions, depending on the specific lakes that were studied and the methodology that was employed. A recent study (Willis 2009) concluded that growth and condition of small mouth bass does not decline when alewives are present, and, for three of the four lakes in the study, diet overlap was low.

In 2008, the Woodland Dam was reopened, allowing alewives access to 2 percent of historic habitat. In 2009, the Maine Legislature's Joint Standing Committee on Marine Resources attempted to reverse the 1995 ruling which closed the fishways at the Woodland and Grand Falls dams, but were stymied as a result of what some observers describe as complex political maneuvering. Consequently, in 2010, the Committee implored the International Joint Commission (IJC), established by the Boundary Waters Treaty in 1909, which authorizes use of lakes and rivers along the border, i.e., the agency that had jurisdiction over the St. Croix, to "resolve the political and diplomatic impasse" and to ensure fish passage, open spawning habitat, and restore the alewife resource of the St. Croix. Over fifty groups, including the Natural Resource Council of Maine, Maine Lobstermen, Maine Rivers, the Atlantic Salmon Federation, representatives of First Nation and tribal interests, and various conservation, sporting, and fishing organizations from the United States and Canada, petitioned the IJC to reopen the alewife run and restore habitat potential for millions of alewives. Some agencies were on board because they envisioned the positive impact of larger numbers of alewives in increasing the food supply for groundfish such as cod, haddock, pollock, and halibut, and the potential to open the run for harvest of lobster bait.

The IJC took up the challenge and proposed a compromise adaptive management plan that would open fishways but exclude alewife from important bass fishing grounds: a delicate ecological balance. Environmentalists weighed in by maintaining that it is a tricky situation when the restoration of a native species is linked to maintenance of an introduced species. Furthermore, because Canadian law prohibits any actions that can harm a native species, the compromise that the IJC sought was not workable because it protects an introduced species at the expense of a native one.

The IJC proposal was in limbo. Year after year, there were missed opportunities for actionable, low-cost methods (the fish ladders were already in place) to restore alewives to the St. Croix, a river that has the potential to serve as the largest river herring restoration project in Maine. But finally, in

June 2013, a victory for alewives and their supporters was achieved when the Maine Legislature voted to approve LD72, a bill that mandates the opening of St. Croix fishways. The sponsorship of the Passamaquaddy Tribe was a key piece of this important reversal of the 1995 law.

New Hampshire

In New Hampshire, dams are being removed, fish passage is being provided for spawners as well as juveniles, and projects have been initiated for the assessment of historical population sizes so that target numbers can be established. Dams interrupt the flow of all major coastal rivers (Cocheco, Exeter, Oyster, Lamprey, Taylor, and Ninnicut) as well as the tributaries of the Connecticut River. In addition to the dams, it appears that the increase in development along the New Hampshire coast and its rivers has severely compromised water quality. The most recent detriments to river herring are increased water draw downs by local communities and the resulting decrease or lack of oxygen, and higher water temperatures. Heavy spring rains often make existing fish ladders impassable. Some have even blamed increases in upstream pesticide use by agricultural interests as contributing to declining water quality.

Because of myriad causes of river decline, several New Hampshire restoration plans are taking a holistic approach to restoration, which involves dam removal (when feasible), fish ladder installation, biological and chemical monitoring studies, river corridor protection, stream crossing improvements, and stormwater runoff mitigation.

Massachusetts

Massachusetts has close to 150 fishways in approximately 100 herring runs, scattered throughout the east and southeast areas of the state. If you have ever driven on Interstates 93, 95 and/or 195, you have driven over most of these herring runs. The seemingly large number of fishways has not helped with the recovery of the state's decimated river herring populations. Some of the fishways don't work very well under certain environmental conditions such as heavy spring rains while others are in need of repair or replacement.

Most of the planned restoration projects in Massachusetts seem small-scale compared to those in Maine. But small-scale does not necessarily translate into projects that are easier to design or faster to complete. As in projects in the Penobscot and Kennebec watersheds in Maine, many stakeholder interests must be considered and partnerships must be built and nurtured. Even with well-intended and well-planned projects, there is always the question about where the funding is coming from.

Alewives Anonymous is group of river herring advocates from southeast Massachusetts, where the Sippican and Mattapoisett Rivers flow through the towns of Rochester, Marion, and Mattapoisett to Buzzards Bay. Over the years, trapping and trucking has been used to supplement the runs, which were important moneymakers for the three towns. For example, in 1883, river herring harvested in weirs on the Mattapoisett River earned the town of Mattapoisett $3,214, which just about covered the annual cost of operating the public school ($3,516) (Alewives Anonymous). The group describes itself as "The Herring Helpers" and has been instrumental in clearing the rivers of storm debris, installing fish ladders and fish counters, and replacing culverts. Each of the rivers has several dams, and hence several fish ladders to assure passage around the dams. On the Sippican, there is a fish ladder at the Hathaway Pond dam, a former mill site, while the final 30 feet of the Sippican's route to the Leonard's Pond spawning area is a tunnel under a roadway. Due to degradation of the riparian habitat and poor water quality, the Sippican River needed a major infusion of help, more than could be provided by a grassroots organization like Alewives Anonymous. A long-term project to restore the Sippican has been initiated by the Coalition for Buzzards Bay, which entered an agreement, in December 2011, with Beaton's, Inc., which uses Hathaway Pond as a water source for its 55-acre cranberry bog. The coalition purchased the dam and 10 acres of adjoining land. The agreement provides for Beaton's continued use of the dam for its cranberry operation until a new reservoir system, one that is permanent and sustainable, can be planned and constructed. In the interim, a new fish ladder will be put in place to allow river herring passage to the pond. It is estimated that it may take 10 years or more before this phase of the project is completed, at which time the dam will be removed.

The Coalition for Buzzards Bay was also instrumental in the rehabilitation of the Acushnet River, which flows 8.2 miles from its headwaters in Freetown, through Acushnet, New Bedford, and Fairhaven into Buzzards Bay. During

2007–2008, along with federal partners such as NOAA, and many other groups, the organization modified three impediments on the river to allow fish passage. Property was acquired, dams were removed or modified, channels were altered, and banks were reconstructed and stabilized. The Sawmill Dam was removed and a nature-like fishway was constructed in its place; the Hamlin Street dam now has a stone weir/pool fishway, and the impediment at the New Bedford Reservoir has a Denil-type fishway. Projects are planned to restore or replace "living resources" in the upland riparian zone.

The cranberry grower A. D. Makepeace is the largest private landowner in Massachusetts and largest supplier of the crop to Ocean Spray. Business has been booming. In 2010, after years of planning and $3 million the MA Division of Fisheries and Wildlife acquired the 245-acre Century Bog, contiguous with other protected parcels, in Wareham and Plymouth from Makepeace. The purchase marks the beginning of the major restoration of the Red Brook watershed, which includes some flow through cranberry bogs (with impediments to river herring migration). The Red Brook, like the Sippican, Mattapoisett, and Acushnet Rivers, also flows into Buzzards Bay. The ecological restoration project planned for the area will protect a native sea-run brook trout stream, and conserve and restore habitat for river herring and other rare or threatened species.

The North Shore is an area of Massachusetts between Boston and the New Hampshire border. The largest river, the Merrimac, is the site of a hydraulic fish lift which helps with fish passage at a large hydropower dam, but the smaller coastal rivers, including the Parker, Little, Saugus, Ipswich, Mill, and Essex Rivers, have many small dams which powered industries in the 1800s but are no longer functional. On the Ipswich alone, there are more than 70 dams and 500 road-stream crossings. Many of the dams are in poor condition and could cause flooding and other problems should they breach. Thus, there is a growing movement to negotiate with dam owners to plan for their removal. In the interim, fishways have been repaired or replaced to keep the runs active. On some of the North Shore rivers, restoration efforts are attempting to naturalize long stretches of river by removing dams and simulating the original condition of the waterway.

Flowing through Boston, the Charles River has been extensively modified. Its salt marshes were filled in and the Back Bay was created. Most of the 80-mile course of the river is influenced by human development, from Boston itself through the suburban landscapes that sprawl along and within the watershed. With about twenty dams and industrial enterprises along its banks, the Charles became one of the most polluted rivers in New England. Although popular for sailing and sculling, it was not very clean until a major initiative was undertaken to make the river "swimmable." The Charles River Watershed Association has worked to clean and restore the Charles, as well as to improve fish passage, and their efforts are paying off. Intrepid river herring, mostly bluebacks, navigate the locks at the mouth of the river and still make their way up the Charles. However, only 20 percent of the run passes beyond the Watertown Dam, and many of the bluebacks spawn right below it. Efforts to restore fish passage west of the Watertown Dam are concentrated on improving fish ladders or installing new ladders so that the herring can navigate further up the river to historic spawning grounds. On spring days, fisherman in waders can be observed near the base of the fish ladder. This is a sign that the blueback herring are running and their natural fish predators are not far behind.

The Lower Neponset River project has identified seventeen miles of river for clean-up and fish passage restoration. The Neponset forms the southern boundary of the city of Boston, originating near the New England Patriots' stadium in Foxboro and emptying into Cape Cod Bay near the large, painted LNG tank on I-93. Today, most of the river is inaccessible to the public, but restoration plans call for increased recreational opportunities as the river is cleaned up. The multifaceted project calls for dam removals, creation of nature-like fishways, and removal of industrial contaminants, like PCBs, from river sediments. This is sure to be a multimillion-dollar project, requiring multiagency cooperation. Reports indicate that progress is slowly being made on all fronts of this complex undertaking.

Although there are no dams on Stony Brook in Brewster, even "The Run" that John Hay made famous needed some help. As the Stony Brook run meanders up from Paine's Creek, it passes through narrow culverts, one of which runs under Route 6A, through property owned by the Cape Cod Museum of Natural History. With the solidification of partnerships among the town of Brewster and other agencies, and funding from NOAA and other sources, the culvert was replaced by an 18-foot box culvert to increase tidal flow, restore 20 acres of degraded salt marsh, and enhance the passage of river herring to their spawning ponds. Plans are being developed to replace other culverts along Stony Brook.

The largest river/estuary/salt marsh restoration project in Massachusetts extends over the towns of Wellfleet and Truro and part of the Cape Cod National Seashore. Close to the mouth of the Herring River, a dike, installed in 1909, greatly restricts tidal flow into the river. The area above the dike has been transformed from a salt marsh of over one thousand acres, into a freshwater marsh and freshwater ecosystem. Only ten acres of the original salt marsh remain. The dike was apparently erected for mosquito control (it failed miserably in this regard by preventing access of aquatic mosquito predators into mosquito breeding grounds), and perhaps to reclaim developable land, thus increasing Wellfleet's appeal as a tourist destination. The structure shrank the effective mouth of the river from 122 meters (400 feet) to a bit less them 2 meters (6 feet). Two of the dike's sluice gates are completely closed; the third is slightly open. Compared to pre-dike conditions, very little water is exchanged between the freshwater and saltwater sides of the structure; the water is often anoxic and acidic, and fish kills are periodically documented. It's a miracle that river herring have figured out how to pass through the rushing downstream flow of the partly open sluice gate. The fact that herring aggregate in front of the dike before attempting to pass through the gate makes this spot a magnet for predators such as striped bass and marks the area as a prime spring fishing spot.

The idea to remove the dike first surfaced in the 1970s when the dike needed extensive repairs, but the issues were not well understood and the motion did not pass at Wellfleet's annual town meeting. Despite the perceived ecological benefits of a proposed restoration, there were many concerns about the potential for increased rise of the river with subsequent flooding of private properties and areas of an adjacent nine-hole golf course as well as intrusion of salt water into private wells. Some of the issues that needed to be addressed to move the project forward included identification of potential archeologically important sites, modeling parameters such as tidal flow, sediment transport, and water levels, identifying properties that could be potentially flooded, and working with property owners to arrive at some type of agreement. Over a decade of scientific studies, meetings with stakeholders, negotiations, public outreach, and education have brought the project close to fruition. A partnership consisting of over two dozen members, including federal, state, municipal, NGO, and private agencies is planning to restore the Herring River and 1,100 acres of its salt marsh wetland. This multimillion-dollar, multi-decade project is expected to return the river to a more healthy state and increase the number of river herring that I and my colleagues count each spring, but, as of 2013, several additional years of permitting and fund-raising are necessary to make this undertaking shovel-ready.

Rhode Island

Rhode Island, with its many embayments, has coastline disproportionate to its small size and watersheds that blanket the state. In southern Rhode Island, with the support of local watershed associations, engineers and contractors have been busy designing projects, removing dams, modifying river channels, and installing fishways. The Pawcatuck River system meanders from a spawning habitat at Worden Pond to its outlet in Westerly. Two dams have been removed; the Kenyon and Lower Shannock Falls Dams (site of a Native American fishing falls before the dam was put in place), and in their place, nature-like fishways were created. However, the third impediment that must be addressed, the Horseshoe Falls Dam, presents a number of challenges to fish passage because removal of this dam is not feasible. The picturesque U-shaped dam, built with stone and mortar, is a historic structure in its own right; it sits in the heart of a historic district and is under private ownership. So, rather than removing the dam, plans are being formulated to install a fish ladder/eel pass near the structure. The design and permitting of the fish ladder is under scrutiny of the state and town historic commissions as well as adjacent property owners. When the project is finalized, three miles of river will be open to river herring.

The Pawtuxet River, flowing through many urban areas and restricted by 140 dams, forms the largest of Rhode Island's watersheds. The 200-year-old Pawtuxet Falls dam sat at the site of the original settlement of Pawtuxet Village, around 1638 and was the first impediment that anadromous fish encountered in their journey from Narragansett Bay. Although the dam had historic significance, it was no longer functional and in need of repair. Removal of this dam was a priority for the Narragansett Bay Estuary Program, and was incorporated into the planning of the Pawtuxet River Anadromous Fish Restoration Project. In August 2011, the heavy equipment appeared and the jackhammers started in on the largest dam removal project in the state. In planning for the removal of the dam, provisions were made for the river to flow over natural bedrock, creating a nature-like fishway. As part of the project, a low-flow channel was cut in, allowing fish to emigrate during dry periods when the water tends to be low.

Connecticut

Steve Gephard supervises the diadromous fish program for the Connecticut Department of Environmental Protection. Gephard explains that Connecticut was the first Atlantic Coast state to recognize the precipitous decline of its river herring stocks and also, in 2002, became the first state to impose a moratorium on the taking of river herring. Conservation and restoration efforts are under way to restore runs by providing access to spawning habitat, to get back to a 1986 operative baseline. Some increases in runs have been observed in several locations where fishways have improved fish passage at dam sites, including the Mianus River in Greenwich, Queach Brook in Branford, and Pattagansett River in East Lyme. Some runs have been restored by restocking; Latimer Brook in East Lyme is the recipient of pre-spawn adult alewives from Bride Brook. Other projects include opening up streams and improving culverts. Where it is feasible, dams are also being removed.

One of Connecticut's most significant urban dam removal and comprehensive restoration projects was initiated in Stamford in 2009 with the removal of two dams, one of which, the Mill Pond Dam, was originally constructed in 1647 as a grist mill and later rebuilt in the 1920s to impound water on the Mill River by the Diamond Ice Company. Removal of the dams, coupled with dredging and removal of contaminants, has made 4.5 miles of river available to migrating fish. After the dams were removed, the unfettered Mill River herring run was given a boost by the introduction of four hundred alewives, trucked in from the Bride Brook run in East Lyme.

Restoration of runs on the Connecticut River and its tributaries, which historically supported a large run of blueback herring, involves efforts from Connecticut, Massachusetts, New Hampshire, and Vermont. When feasible, dams are being removed, but most restoration efforts involve the construction of some type of fish passage. As with the Stamford project, the recovery is accelerated by restocking programs. For example, the Massachusetts Division of Fisheries and Wildlife installed a fish passage for upstream and downstream migration in 1996 at a dam on the Westfield River in West Springfield, opening 14 miles of river, but it was not being used by blueback herring. In 2000, a trap and transfer program was initiated in which blueback herring were collected from the Connecticut River by electrofishing, a process in which fish are electrically shocked and float to the surface (The fish can be scooped up and will usually recover when they are released back into the water.). The shocked fish were released above the fishway where they would have access to spawning habitat, with the hope that they, and their progeny, would return to this portion of the Westfield River.

New York

In New York, there are many small coastal rivers that flow into Long Island Sound, Peconic Bay, and the south coast of Long Island. As in other locations, partnerships among communities, schools, industry, government, private landholders, and nonprofit agencies seek to restore fish passage by removal of dams and construction of fish ladders. Although these projects are smaller in scope than those of Maine's great rivers or the Herring River in Wellfleet, they are numerous. Many of the projects are in the planning and permitting stage, and several others have been completed and are already making a difference to spring migration and spawning. These efforts are vitally important in restoring herring runs, some of which are extirpated (locally extinct), while others are remnants of their former glory.

Run Revitalization

It is apparent from considering the sampling of case studies for restoring, renewing, enhancing, or rehabilitating river herring runs that each project is very complex and also very specific to its location and the overall goals that stakeholders envision. Just as "one size fits all" does not apply to fishway design, the motto also is not applicable to restoration projects. The planning, permitting, monitoring, management, and expense of river and river herring run restoration projects mandates that many partners are involved. It sometimes takes a monumental effort to identify stakeholders and get them all into the same room. Although many of the restoration projects appear to be creeping along, there are signs that some plans are moving forward at a more rapid rate and there are even some true success stories. The question remains: Will these projects actually make a difference to the recovery of the species? If the number of individuals has declined to a trickle of their former level, and the number of stocks is so low, will river herring be able to rebound even if their habitat has been given back to them?

CHAPTER 11
A Call to Action

If the numbers of river herring have declined to a trickle of their former levels, and none of the efforts to boost their populations has so far succeeded, will river herring ever be able to rebound? Those who are working on restoration projects and those who count herring each spring are optimistic that we will see a partial rebound of river herring in our northeast runs. This will not necessarily mean complete restoration; that would be impossible because historic runs that broiled and roiled each spring did not have today's level of anthropogenic influence. Many New Englanders live close to the water, so our homes, activities, and industries all have an impact on our rivers, lakes, ponds, streams, estuaries, and the ocean itself. At best, we may be able to achieve a level of restoration in which the habitats for river herring are functioning as sustaining ecosystems. River herring are underdogs; they fight their way upstream with a tremendous expense of energy; the least we can do is expend some of our energy to rally to their cause. This will take efforts on many levels. And we must remember: it's not just about the fish. In all cases, it's also about their habitat and the membership and balance of the ecosystems which are required to sustain their populations.

Federal level: At the federal level, it is up to our fisheries managers to protect river herring by managing the Atlantic fisheries in a manner that prevents bycatch of river herring. Amendment 14 to the Atlantic Mackerel, Squid and Butterfish Fishery Management Plan was a step in the right direction but cannot be the end of the effort. Amendment 5 to the Atlantic Herring Fisheries Management Plan was a dismal failure with respect to protection of river herring. NOAA fisheries has also been remiss in acting on the scientific data which point to severe declines in the species and offering river herring much-needed protection.

States and municipalities (inland fisheries management) must do their share. States and towns that want to open or continue harvest of river herring must develop sustainable fishery plans and submit them to the appropriate fishery managers for approval. These plans must include sufficient data to assure the sustainability of the runs. In addition, the appropriate agencies should streamline the management and planning of restoration projects. Sometimes it takes a human generation before any river restoration project can be realized. By cutting down on some of the unnecessary red tape, it may be possible to fast-track the permitting processes and get these projects off the ground and running.

Environmental organizations and NGOs such as American Rivers and The Nature Conservancy are advocates for dam removal and serve as brokers for fish passage by providing assistance, advocacy, and exploration of funding sources for projects. Without the experience, assistance, and direction of these large organizations with a broad mission, many well-intentioned site-specific restoration projects would be floundering.

Science: As well as basic science research on the natural history and physiology of anadromous fishes, the scientific community is providing some leadership and data to support restoration projects and management decisions. For example, there is an ongoing project to define the river herring populations and assess the genetic variation within and among the various stocks. These studies attempt to define a "stock" so that there is information about loss of specific stocks during bycatch events, indications about how much straying is occurring during emigration to natal spawning grounds, and the potential to repopulate stocks once they have been lost or depleted. This type of information will be critical to determine what is happening to individual runs. The report produced by the NMFS River Herring Stock Structure Working Group suggests the genetic differentiation of five major stocks for alewife and four for blueback herring, and "that the major river drainage is the appropriate level of management" for river herring (Palkovacs and Gephard 2012).

The Sustainable Science Initiative (SSI), led by a consortium of faculty and students from Bates College, Bowdoin College, and the University of Southern Maine, supports many projects, including one with a focus on river herring in the Kennebec and Androscoggin Rivers. The group is examining the quality and quantity of river herring spawning habitat, as well as the economic benefits of river restoration as it impacts the river herring fishery. An additional focus is the potential for river herring to affect groundfish populations closer to shore. At the University of New Hampshire, Jamie Cournane and Christopher Glass worked with the National Fish and Wildlife Foundation to use long-term catch data to determine the historic size of river herring populations. This estimate, along with up-to-date stock assessments, will help to inform fisheries managers about restoration "targets." The University of Massachusetts (Dartmouth) is working with the Massachusetts DMF to produce a fine-scale map of river herring bycatch events to further delineate and predict river herring "hot spots." This information may also be used to develop a bycatch avoidance system for the Atlantic herring fleet. For example, the fleet may be required to avoid the identified hot spots, and protocols can be developed to allow vessels to communicate with one another when they find river herring in their targeted harvest.

Local action: On a cold, damp day in March 2011, over thirty individuals attended the Herring Count Workshop, hosted by Alex Mansfield, Ecology Program Director at Jones River Landing in Kingston, Massachusetts. The purpose of the workshop was to standardize herring count data in the state's municipalities so that they can be integrated into a single online database, and to incorporate the data into the Massachusetts Bays Project on climate change. Plans were formulated to install data loggers to measure temperature, rainfall, and water flow in many of the locations where herring are counted. It was the first time all Massachusetts river herring stakeholders were gathered together in the same place. The workshop also afforded herring wardens, watershed associations, and interested individuals the opportunity to share tips (such as not to wear bright yellow or red when counting). This was also the venue in which Wellfleet Herring Warden Jeff Hughes introduced his concept for a River Herring Network. Hughes obtained a modest grant and the support of the Cape Cod Commercial Fishermen's Alliance (formerly named the Cape Cod Commercial Hook Fishermen's Association) for the network. The network includes a website (www.riverherringnetwork.com) at which river herring wardens throughout New England can share ideas, observations, and management tips, recruit volunteers, share count data, and devise a manual for best management practices. There was a great deal of enthusiasm and optimism from those who attended the workshop, as well as a commitment to do all that can possibly be done to enhance and restore the inland habitat for river herring. The website is up and running and already has a wealth of information for herring wardens, count volunteers, and interested members of the public. I have attended meetings of the River Herring Network and have found them to be an invaluable source for learning about ongoing projects, results of scientific studies, conservation initiatives, and state and federal guidelines. In addition, it is a terrific venue for informal discussions among attendees, who include count volunteers, town employees, state officials, and other stakeholders.

Groundswell of public pressure: The general public is becoming more cognizant of the fact that habitat restoration helps all species in the ecosystem, including humans. They are also more aware of the importance of forage fish in aquatic ecosystems. H. Bruce Franklin's *The Most Important Fish in the Sea: Menhaden and America* has driven home the message that ecosystems depend on their forage base. The Herring Alliance and the Cape Cod Commercial Fishermen's Alliance have been instrumental in highlighting the issue of protection of forage fishes and fishery resources, as well as the issue of bycatch in the Atlantic herring fishery. With growing awareness of the overall role of fishes like menhaden, Atlantic herring, and river herring in the health of aquatic ecosystems, many environmental organizations, sports fishermen's groups, watershed organizations, and grassroots community associations are writing and meeting with their mayors, senators, and representatives about their concerns regarding the ecological depletion of forage species in aquatic food webs.

Endangered

On August 1, 2011, the Natural Resources Defense Council (NRDC) filed a 98-page petition to the U.S. Department of Commerce and NOAA, to "List Alewife (*Alosa pseudoharengus*) and Blueback Herring (*Alosa aestivalis*) as Threatened Species and to Designate Critical Habitat." The nonprofit NRDC, a collaborative of lawyers, policymakers, and scientists, takes on many environmental issues as part of its mission "To safeguard the Earth;

its people, its plants and animals and the natural systems on which all life depends." Reviving the world's oceans is a top priority. The NRDC petition also includes an alternative; if the river herring are not listed as threatened according to criteria in the Endangered Species Act, NRDC called for the designation of specific segments of the populations as Distinct Population Segments (DPSs) and the listing of river herring in each DPS as threatened. The regions to be so designated are described as discrete on the basis of separation due to physical, physiological, ecological, or behavioral factors. The genetic analysis of river herring and stock structure studies would help to define these designations if river herring from each of the segments exhibited considerable genetic or morphological substructure.

Once the petition was received, the National Marine Fisheries Service was required to make a determination whether the information in the petition "presents substantial scientific or commercial information indicating that the petition action might be warranted," and respond within ninety days. In other words, the petition was a first step: the federal government was not required to make a decision on a potential "threatened" listing for river herring; it simply had to conclude that the petition contained enough evidence for the change in designation to be considered. The petition itself is an encyclopedia of information about river herring and describes scientific and fisheries data, river by river, which support the petition. On November 1, 2011, the NMFS complied and announced that the petition contained sufficient information and will be subject to further review.

Essentially, a "threatened" classification implies that if nothing is done to change the status quo, a species can expect to decline to "endangered" status within a specified time frame, which, in the case of anadromous fishes is usually set for one hundred years. The process of reviewing scientific data to make the final determination was expected to take at least a year, but the review dragged on beyond the deadline without a final decision. If the designation was changed from "Species of Concern" to "Threatened," both species would have additional conservation protection, either as aggregates or within DPSs. A potential change in the federal listing of river herring may be a blessing for river herring, but it has the potential to shut down many important northeast and Mid-Atlantic fisheries as well as inland fisheries. It's no wonder that The Alewife Harvesters of Maine opposed the listing.

On August 9, 2013, NOAA made an announcement regarding the decision. The proposed listing of river herring as either "threatened" or "endangered" under the Endangered Species Act was deemed "not warranted" at this time. Instead, river herring supporters were encouraged to continue to collect data and work toward conserving the fish and their habitat—activities that have been in place for decades without demonstrable positive impacts on river herring populations.

Diminishing Returns

Alewives are already extirpated in South Carolina; will other states follow in this trend? Each spring, I will find myself back at the Herring River in Wellfleet, recording environmental data and watching the water for signs of life. Even in this rather pristine run, the trends have been disappointing. With "Visual Count" software, developed by the Massachusetts Division of Marine Fisheries, the herring counts on all runs monitored by volunteer counts have been recalibrated. Our first Wellfleet census was 17,000 in 2009, which dipped to 12,500 in 2010, and 7,700 in 2011, was calculated at 12,000 in 2012, and rose to 25,000 in 2013. Time will tell the meaning of these numbers. Experts agree that we can't implicate a single factor as cause for the continuing decline of this once iconic small fish, a harbinger of spring and part of the cultural heritage of many coastal and river cities and towns. The cumulative effect of diverse negative impacts are taking their toll. Attention to one area, whether it be exploitation in the inland fisheries, bycatch at sea, passage impediments, pollution, or others, cannot be expected to fix the problem. Given the nature of the fish's life history, the probability of a river herring egg becoming fertilized and developing into an adult is very, very low, so there is a requirement for a large number of fish to sustain the population. The fish must also be adequately dispersed along the coast, rather than limited to a few distinct drainages, in order to ensure sufficient genetic diversity and thus the ability of the species to respond in a positive way to changing environmental conditions. The numbers of fish that are documented in particular runs do not come close to historic levels and, in the worst case scenario, may represent remnant populations. The populations of alewives and blueback herring may already be stretched to the limit and at their tipping point. They have arrived there as a result of multiple factors with cumulative and synergistic negative impacts.

Could river herring be the next modern extinction? The passenger pigeon disappeared at the beginning of the twentieth century. Like river herring, the passenger pigeon was a migrating species, treating onlookers to spectacular annual displays. They were once the most common birds in North America, constituting 25–40 percent of the total bird population in the United States before European colonization. However, it didn't take very long before loss of habitat and overhunting decimated them. Their precipitous decline made it impossible for this species to persist because large numbers of individuals were needed for the species to recover. Once a serious dent was made in the population, there was no way out. The same scenario may be playing out for river herring. With the exception of fishermen who valued them for bait, river herring became a neglected part of their ecosystems during the last part of the twentieth century. Now that we have started to pay attention, and with concerted effort, river herring might be rescued from the brink of extinction. Only our continued vigilance and the passage of time will tell whether all our belated efforts have made any difference.

APPENDIX

See How They Run

Although there are many river herring runs throughout the Northeast, very few of them offer good opportunities to see river herring. Some are on large rivers, like the Hudson, Connecticut, Penobscot, and Kennebec, while others are on private property. Some of the best locations to see river herring are at well-maintained fish ladders where efforts have been made to provide a prospect for viewing. There is a wide window of time during spring when you might see migrating river herring in the Northeast, generally from mid-March to the end of May. The timing also depends on whether the run you wish to visit consists of alewives or blueback herring or both. Generally, the farther south, the earlier the fish will start running. And, of course, there are no guarantees. River herring might be there one day, but not the next. It has sometimes taken me several trips to a run before I have seen fish, and on certain runs I've never had success. Talk to your local herring warden or herring inspector; he or she is most likely to have a good sense of the timing of the runs in your area and the local conditions which are the most conducive to a fruitful trip.

MAINE

The **Damariscotta Mills** fishway (see fig. 10.1) is tailor-made for visitors. In 2007, extensive repairs were initiated on the fishway, and the project is expected to be finished in 2014. The restoration of the fishway is under the supervision of the towns of Nobleboro and Newcastle in collaboration with the Nobleboro Historical Society. The new fishway design offers resting pools, separated by stone and mortar weirs, and provides walkways for visitors. Check out the website, www.damariscottamills.org, for information about the spring and fall fish festivals, photos of the alewife harvest, and some wonderful historic and modern photos that depict how alewives are strung, brined, and smoked.

Location: Damariscotta Mills Road. Nobleboro, Maine.

MASSACHUSETTS

An overview of accessible river herring runs in Massachusetts is available at www.mass.gov/eea. This brochure, titled "A Guide to Viewing River Herring in Coastal Massachusetts," indicates the location of herring runs on a map of the southern and eastern sections of the state. In addition, there is contact information for various watershed groups on the North Shore, Boston area, South Shore, and Cape Cod. Nine coastal Massachusetts fishways are described, and their locations are indicated. Some of the viewing areas I recommend overlap with those described in the pamphlet.

Massachusetts North Shore

North Shore watersheds and fishways are described in a Massachusetts Division of Fisheries and Wildlife Technical Report (Reback et al., 2004b) available at www.mass.gov/eea. Most of the fish ladders on the rivers north of Boston are inaccessible to the public.

Medford/Somerville: Although the dam and fish ladder are not accessible to the public, Mystic River herring are celebrated annually. The Mystic River Herring Run and Paddle is sponsored by the Mystic River Watershed Association, which celebrates the river and its historic herring run. Held each May, the event starts at the Blessing of the Bay Boathouse in Somerville and draws runners, kayakers, and even stand-up paddleboarders. Contestants are urged to "Run, walk or paddle for the fish!"

Newbury: The Woolen Mills fishway on the Parker River is relatively easy to find, but there is no parking at the site. You may have to park your vehicle in an open area on the shoulder of the road and walk to the fishway.
Location: Central Street and Parker River (near Orchard Street intersection)

Watertown: The Watertown Dam and fishway are in the River Front Park on the Charles River. Parking can be found on the local streets. The best place to see blueback herring is at the base of the dam where they gather to spawn.
Location: California Street and River Front Park, Watertown.

Massachusetts South Shore

South Shore dams and fishways are described in a Massachusetts Division of Fisheries and Wildlife Technical Report (Reback et al., 2004a) available at www.mass.gov/eea.

Bournedale: There are two viewing spots on the Monument River near the Cape Cod Canal. One area is at the Herring Run Recreation Area on the north side of the canal, while the other, more scenic area is at Carter Beal Conservation Area just north of Route 6 (Scenic Highway). Ample parking is available at both locations.
Locations: Route 6 at Army Corps of Engineers Herring Run Recreation Center and off Bournedale Road, across Route 6 from the Visitor Center.

Brewster: The Stony Brook run (see fig. 1.5) is easily accessible, and there is parking for a few cars. This is one of the most famous river herring runs in the Northeast and has many visitors, not only during spring, but also throughout the year. The small park along the banks offers areas for strolling and picnics.
Location: Stony Brook Road near Setucket, across from the old grist mill.

East Weymouth: Around Jackson Square much of the Back River flows through culverts, under the roads, so the fishway at the pocket park is the best viewing area. Parking can be found on the street, although you may have to drive around a bit to find a spot.
Location: Near intersection of High, Water, and Pleasant Streets.

Holyoke: The hydraulic fishlift on the Connecticut River presents an opportunity to see blueback herring and shad though the viewing windows. The viewing area is open from early May through June, generally from Wednesday through Sunday from 9 a.m. to 5 p.m. Information about the lift, dates of operation, public viewing, and driving directions can be found at www.hged.com, under Robert Barrett Fishway.

Kingston: This dam and fishway on the Jones River are easily accessible from Elm Street.
Location: Elm Street between Brook Street (Route 80) and Main Street (Route 106).

Marstons Mills: The fishway on the Marstons Mills River near Mill Pond is easily accessible for viewing, but be careful because it is next to a busy intersection. There is limited parking.
Location: On Route 149 at intersection with Route 28.

Mashpee: This fishway is next to the Wampanoag Indian Museum and there is ample parking.
Location: Route 130 near the intersection of Route 28.

Mattapoisett: The Mattapoisett River empties into Buzzards Bay. The fish ladder in the town of Mattapoisett is adjacent to a cave-like area close to Route 6 where river herring were processed. There is room for a car or two to park near the fishway.
Location: Route 6 in Mattapoisett.

Middleborough: This herring run, on the Nemasket River, is the largest in the state, and Nemasket herring have been used to restock other runs. Off-street parking is available near the fishway (see fig. 10.2) at Oliver Mills Park.
Location: Wareham Street (off Route 105) in downtown Middleborough.

Plymouth: The vicinity of the historic Jenney Grist Mill (see fig. 7.1), in downtown Plymouth, is a delightful setting to view river herring. Town Brook meanders through the area, and it is easy to see the schools of alewives from the banks. There is also a fishway near the mill itself. A large parking lot accommodates visitors to the historic area.
Location: Spring Lane.

Wareham: The fish ladder at the Mill Pond Dam is a good viewing spot for Agawam River herring. There is a large parking lot adjacent to the fishway at the Elks Club.
Location: Elks Club parking area; Route 6/28.

Rhode Island

North Kingstown / Saunderstown: This small run is on Gilbert Stuart Brook near the Gilbert Stuart Birthplace and Museum. Ample parking.
Location: Gilbert Stuart Road.

Connecticut

Greenwich: The Mianus River dam and fish ladder provide a great opportunity to see a fishway and eel pass close up during one of the spring open house events (see fig. 1.4). The aluminum walkway that runs parallel to the fish ladder allows visitors access to the top of the ladder where the electronic fish counters are in place. Call the Greenwich Conservation Commission at 203-622-6461 for information about dates for open house events at the fishway. Off-street parking is available.
Location: 125 River Road Extension, Cos Cob.

Windsor: Viewing windows allow visitors to watch passing fishes on the Farmington River at the Rainbow Fishway. You may not see any river herring, but if you live in northern Connecticut or south central Massachusetts, this may be your closest opportunity to see anadromous fishes. Parking is available at the site. For information and driving directions, see www.fws.gov/r5crc, under Where to See Fish.

Works Cited

Ames, Edward P. 2004. "Atlantic cod stock structure in the Gulf of Maine." *Fisheries* 29: 10–28.

Ampela, Kristen. 2009. "The diet and foraging ecology of grey seals (*Halichoerus grypus*) in United States waters." PhD diss., City University of New York.

ASMFC. 2009. Amendment 2 to the Interstate Fishery Management Plan for Shad and River Herring (River Herring Management). www.asmfc.org.

Becker, James. 2009. "Commercial catch sampling and bycatch survey of two pelagic fisheries." Atlantic Coastal Cooperative Statistics Program. www.nero.noaa.gov.

Belding, David L. 1920. "A report upon the alewife fisheries of Massachusetts." Commonwealth of Massachusetts, Department of Conservation. Division of Fisheries and Game.

Bozeman, E. L., Jr., and M. J. Van Den Avyle. 1989. "Species profiles: life histories and environmental requirements of coastal fishes and invertebrates (South Atlantic)—alewife and blueback herring." U.S. Fish and Wildlife Service Biological Report no. 82 (11.111). U.S. Army Corps of Engineers, TR EL-82-4.

Buckley, Betty, and Scott W. Nixon. 2001. "An historical assessment of anadromous fish in the Blackstone River." Final report to the Narragansett Bay Estuary Program, the Blackstone River Valley National Heritage Corridor Commission, and Trout Unlimited. University of Rhode Island, Narragansett, Rhode Island.

Cape Cod Cranberry Growers Association. 2004. "Advisory for anadromous fish passage." www.cranberries.org.

Cavallo, Orlando N. n.d. "The Pembroke herring run and history of the valley and fishery." Pembroke Herring Fisheries Commission. www.pembroke-ma.gov.

Cieri, Matthew, Gary Nelson, and Michael Armstrong. 2008. "Estimates of river herring bycatch in the directed Atlantic herring fishery." Report prepared for the Atlantic States Fisheries Commission.

Connecticut River Coordinator's Office. 2004. Management plan for river herring in the Connecticut River basin. www.fws.gov/R5Crc.

Cournane, Jamie, and Steven Correia. 2010a. "Draft update: identification of river herring hotspots at sea using fisheries dependent and independent datasets." Report prepared for the Atlantic Herring Plan Development Team.

——. 2010b. Spatial management alternatives to address river herring bycatch in the directed Atlantic herring fishery. Report prepared for the Atlantic Herring Plan Development Team.

Davis, Justin P., and Eric T. Schultz. 2009. "Temporal shifts in demography and life history of an anadromous alewife population in Connecticut." *Marine and Coastal Fisheries: Dynamics, Management and Ecosystem Science* 1: 90–106.

Dufour, E., T. O. Hook, W. P. Patterson, and E. S. Rutherford. 2008. "High-resolution isotope analysis of young alewife *Alosa pseudoharengus* otoliths: assessment of temporal resolution and reconstruction of habitat occupancy and thermal history." *Journal of Fish Biology* 73: 2434–51.

Elsdon, Travis S., Brian K. Wells, Steven E. Campana, Bronwyn M. Gillanders, Cynthia M. Jones, Karin E. Limburg, David H. Secor, Simon R. Thorrold, and Benjamin D. Walther. 2008. "Otolith chemistry to describe movement and life-history parameters of fishes: hypotheses, assumptions, limitations and inferences." *Oceanography and Marine Biology: An Annual Review* 46: 297–330.

Fay, C. W., R. J. Neves, and G. B. Pardue. 1983. "Species profiles: life histories and environmental requirements of coastal fishes and invertebrates (Mid-Atlantic): alewife/blueback herring." U.S. Fish and Wildlife Service, Division of Biological Services. FWS/OBS-82/11.9. U.S. Army Corps of Engineers, TR EL-82-4.

Field, G. W. 1914. "Alewife fishery of Massachusetts." *Transactions of the American Fisheries Society* 43: 143–51.

Finch, Robert. 2003. *Special places on Cape Cod and the Islands.* Commonwealth Editions, Beverly, Mass.

Flagg, Lewis N. 2007. "Historical and current distribution and abundance of the anadromous alewife (*Alosa pseudoharengus*) in the St Croix River." A Report to the State of Maine Atlantic Salmon Commission.

Gahagan, Benjamin I., Katie E. Gherard, and Eric T. Schultz. 2010. "Environmental and endogenous factors influencing emigration in juvenile anadromous alewives." *Transactions of the American Fisheries Society* 139: 1069–2010.

Gahagan, Benjamin I., Jason Vokoun, Gregory Whitledge, and Eric T. Schultz. 2012. "Evaluation of otolith microchemistry for identifying natal origin of anadromous river herring in Connecticut." *Marine and Coastal Fisheries: Dynamics, Management, and Ecosystem Science* 4: 358–72.

Gill, Theo. 1901. "Alewives." *Notes and Queries* 104: 451.

Haas-Castro, R. 2006. Status of fishery resources off the northeastern US. www.nefsc.noaa.gov.

Hay, John. 1959. *The Run.* Beacon Press. Boston.

Hightower, Joseph E., Anton M. Wicker, and Keith M. Endres. 1996. "Historical trends in abundance of American shad and river herring in Albemarle Sound, North Carolina." *North American Journal of Fisheries Management* 16: 257–71.

Iafrate, Joseph, and Kenneth Oliveira. 2008. "Factors affecting migration patterns of juvenile river herring in a coastal Massachusetts stream." *Environmental Biology of Fishes* 81: 101–10.

Ihssen, Peter E., G. William Martin, and David W. Rogers. 1992. "Allozyme variation of Great Lakes Alewife, *Alosa pseudoharengus*: genetic differentiation and affinities of a recent invader." *Canadian Journal of Fisheries and Aquatic Sciences* 4: 1770–77.

Jacobs, Donald. 1996. "Herring time." In *Bournedale: the forgotten village,* ed. Michael Bradley. The Bournedale Historical Commission, Bourne, Mass.

Jessop, B. M. 1994."Homing of alewives (*Alosa pseudoharengus*) and blueback herring (*A. aestivalis*) to and within the Saint John River, New Brunswick, as indicated by tagging data." Canadian Technical Report of Fisheries and Aquatic Sciences no. 2015.

Jones, Andrew J., Christopher M. Dalton, Edward S. Stowe, and David M. Post. 2010. "Contribution of declining anadromous fishes to the reproductive investment of a common piscivorous seabird, the double-crested cormorant (*Phalacrocorax auritus*)." *Auk* 127: 696–703.

Josselyn, John. 1988 [1674]. *John Josselyn, colonial traveler: a critical edition of Two Voyages to New England.* Ed. P. J. Lindholdt. University Press of New England, Hanover, N.H.

Kellog, Robert L. 1982. "Temperature requirements for the survival and early development of the anadromous alewife." *Progressive Fish Culturist* 44: 63–73.

Ketola, H. George, Paul R. Bowser, Gregory A. Wooster, Leslie R. Wedge, and Steven S. Hurst. 2000. "Effects of thiamine on reproduction of Atlantic salmon and a new hypothesis for their extirpation in Lake Ontario." *Transactions of the American Fisheries Society* 129: 607–12.

Klauda, Ronald J., and Robert E. Palmer. 1987. "Responses of blueback herring eggs and larvae to pulses of acid and aluminum." *Transactions of the American Fisheries Society* 116: 561–69.

Klauda, Ronald J., Robert E. Palmer, and Michael J. Lenkevich. 1987. "Sensitivity of early stages of blueback herring to moderate acidity and aluminum in soft water." *Estuaries* 10: 44–53.

Kosa, Jarrad T., and Martha Mather. 2001. "Processes contributing to variability in regional patterns of juvenile river herring abundance across small coastal systems." *Transactions of the American Fisheries Society* 130: 600–619.

Limburg, Karin E., and Robert E. Schmidt. 1990. "Patterns of fish spawning in Hudson River tributaries: response to an urban gradient." *Ecology* 71: 1238–45.

Limburg, Karin E., and John R. Waldman. 2009. "Dramatic declines in North Atlantic Diadromous Fishes." *BioScience* 59: 955–65.

Loesch, Joseph G., and William A. Lund Jr. 1997. "A contribution to the life history of the blueback herring, *Alosa aestivalis*." *Transactions of the American Fisheries Society* 106: 583–89.

MacNeill, D. B., and S. B. Brandt. 1990. "Ontogenic shifts in gill-raker morphology and predicted prey capture efficiency of the alewife, *Alosa psueudoharengus.*" *Copeia* 1: 164–71.

Mancera, Juan Miguel, and Stephen D. McCormick. 2007. "Role of prolactin, growth hormone, insulin-like growth factor 1 and cortisol in teleost osmoregulation." In *Fish osmoregulation,* ed. Bernardo Baldisserto, J. M. Mancera Romero, and B. G. Kapoor, 497–515. Science Publishers. Boca Raton, Fla.

McCormick, Stephen D. 2001. "Endocrine control of osmoregulation in teleost fish." *American Zoologist* 41: 781–94.

McCormick, Stephen D., Darren T. Lerner, Michelle Y. Monette, Katherine Nieves-Puigdoller, John T. Kelly, and Bjorn Thrandur Bjornsson. 2009. "Taking it with you when you go: how perturbations to the freshwater environment, including temperature, dams, and contaminants affect marine survival of salmon." *American Fisheries Society Symposium* 69: 195–214.

McDowall, R. M. 1997. "The evolution of diadromy in fishes (revisited) and its place in phylogenetic analysis." *Reviews in Fish Biology and Fisheries* 7: 443–62.

——. 2001. "Anadromy and homing: two life-history traits with adaptive synergies in salmonid fishes?" *Fish and Fisheries* 2: 78–85.

McKenzie, Matthew. 2010. *Clearing the coastline: the nineteenth-century ecological and cultural transformation of Cape Cod.* University Press of New England, Hanover, N.H.

McKeown, Brian A. 1984. *Fish migration.* Timber Press, Portland, Ore.

Messieh, Shoukry N. 1977. "Population structure and biology of alewives (*Alosa pseudoharengus*) and blueback herring (*A. aestivalis*) in the Saint John River, New Brunswick." *Environmental Biology of Fishes* 2: 195–210.

Milstein, Charles B. 1981. "Abundance and distribution of juvenile *Alosa* species off southern New Jersey." *Transactions of the American Fisheries Society* 110: 306–9.

Nelson, Gary A. 2006. "A guide to statistical sampling for the estimation of river herring run size using visual counts." MA Division of Marine Fisheries Technical Report TR-25.

Nestler, John M., R. Andrew Goodwin, Thomas M. Cole, Donald Degan, and

Donald Dennerline. 2002. "Simulating movement patterns of blueback herring in a stratified southern impoundment." *Transactions of the American Fisheries Society* 131: 55–69.

Neves, Richard J. 1981. "Offshore distribution of alewife, *Alosa pseudoharengus*, and blueback herring, *Alosa aestivalis*, along the Atlantic coast." *Fisheries Bulletin* 79: 473–85.

NOAA Fisheries. 2009. "Species of Concern: River herring." www.nmfs.noaa.gov.

Nordeng, H. 1997. "A pheromone hypothesis for homeward migration in anadromous salmonids." *Oikos* 28: 155–59.

O'Connell, Ann M., and Paul L. Angermeier. 1997. "Spawning location and distribution of early life stages of alewife and blueback herring in a Virginia stream." *Estuaries* 20: 779–91.

Palkovacs, Eric P., Kristin B. Dion, David M. Post, and Adalgisa Caccone. 2008. "Independent evolutionary origins of landlocked alewife populations and rapid parallel evolution of phenotypic traits." *Molecular Ecology* 17: 582–97.

Palkovacs, Eric P., and Stephen Gephard. 2012. Report for the NMFS River Herring Stock Structure Working Group. www.nero.noaa.gov.

Palkovacs, Eric, and David M. Post. 2008. "Eco-evolutionary interactions between predators and prey: can predator-induced changes to prey communities feed back to shape predator foraging traits?" *Evolutionary Ecology Research* 10: 699–720.

Pardue, G. B. 1983. Habitat suitability index models: alewife and blueback herring. U.S. Department of the Interior, Fish and Wildlife Service report FWS/OBS-82/10.58.

Pory, John. 1997 [1623]. "John Pory to the Earl of Southampton." In *Three visitors to Plymouth: letters about the pilgrim settlement in New England during its first seven years*. 5–13. Ed. Sydney V. James. Applewood Books. Bedford, Mass.

Post, David M., Eric P. Palkovacs, Erica G. Schielke, and Stanley I. Dodson. 2008. "Intraspecific variation in a predator affects community structure and cascading trophic interactions." *Ecology* 89: 2019–32.

Purinton, Tim, Frances Doyle, and Robert D. Stevenson. 2003. Status of river herring on the North Shore of Massachusetts. http://ipswich-river.org.

Reback, K. E., P. D. Brady, K. E. McLaughlin, and C. G. Milliken. 2004a. "A Survey of Anadromous Fish Passage in Coastal Massachusetts,

Part 3. South Shore." Massachusetts Division of Marine Fisheries Technical Report TR-17, available at www.mass.gov/eea.

——. 2004b. "A Survey of Anadromous Fish Passage in Coastal Massachusetts, Part 4. Boston and North Coastal." Massachusetts Division of Marine Fisheries Technical Report TR-18, available at www.mass.gov/eea.

Reebs, Stephan. 2001. *Fish behavior in the aquarium and in the wild.* Comstock Publishing Associates, Cornell University Press. Ithaca.

Richkus, William Albert. 1974. "Factors influencing the seasonal and daily patterns of alewife (*Alosa pseudoharengus*) migration in a Rhode Island river." *Journal of the Fisheries Research Board of Canada* 31: 1485–97.

Rounsefell, George A., and Louis D. Stringer. 1945. "Restoration and management of the New England alewife fisheries with special reference to Maine." *Transactions of the American Fisheries Society* 73: 394–424.

Simonin, Paul W., Karin E. Limburg, and Leonard S. Machut. 2007. "Bridging the energy gap: anadromous blueback herring feeding in the Hudson and Mohawk rivers, New York." *Transactions of the American Fisheries Society* 136: 1614–21.

Smith, Captain John. 2007 [1630]. "The General Historie of Virginia, New England, and the Summer Isles." In *Writings, with other narratives of Roanoke, Jamestown and the first English settlement of America,* 199–670. Library of America. New York.

Smith, Joseph M., Martha E. Mather, Holly J. Frank, Robert M. Muth, John T. Finn, and Stephen D. McCormick. 2009. "Evaluation of a gastric radio tag insertion technique for anadromous river herring." *North American Journal of Fisheries Management* 29: 367–77.

Smylie, Mike. 2004. *Herring: a history of the silver darlings.* Tempus Publishing. Gloucestershire, U.K.

Solomon, D. J. 1973. "Evidence for pheromone-influenced homing by migrating Atlantic salmon, *Salmo salar* (L.)." *Nature* 244: 231–32.

Thunberg, Bruce E. 1971."Olfaction in parent stream selection by the alewife (*Alosa pseudoharengus*)." *Animal Behaviour* 19: 217–25.

USFWS GOMCP. 2007. "Diadromous fish habitat protection and restoration projects in Maine." Report compiled by the Gulf of Maine Coastal Program, U.S. Fish and Wildlife Service. www.fws.gov.

Visel, Timothy C. 1988. "The Hummers Pond alewife restoration project, Madison, Connecticut." Report submitted to the Madison Land Conservation Trust.

Vladykov, Vadim D. 1936. "Occurrence of three species of anadromous fishes on the Nova Scotian banks during 1935 and 1936." *Copeia* 1936: 168.

Walters, Annika W., Rebecca T. Barnes, and David M. Post. 2009. "Anadromous alewives (*Alsosa pseudoharengus*) contribute marine-derived nutrients to coastal stream food webs." *Canadian Journal of Fisheries and Aquatic Sciences* 66: 439–48.

Walton, Clement J. 1983. "Growth parameters for typical anadromous and dwarf stocks of alewives, *Alosa pseudoharengus*." *Environmental Biology of Fishes* 9: 277–87.

West, Derek C., Annika W. Walters, Stephen Gephard, and David M. Post. 2010. "Nutrient loading by anadromous alewife (*Alosa pseudoharengus*): contemporary patterns and predictions for restoration efforts." *Canadian Journal of Fisheries and Aquatic Sciences* 67: 1211–20.

Willis, Theodore V. 2009. "How policy, politics, and science shaped a 25-year conflict over alewife in the St. Croix River, New Brunswick–Maine." *American Fisheries Society Symposium* 69: 1–19.

Yako, Lisa A., Martha E. Mather, and Francis Juanes. 2002. "Mechanisms for migration of anadromous herring: an ecological basis for effective conservation." *Ecological Applications* 12: 521–34.

Young, Alexander. 1846. *Chronicles of the first planters of the Colony of Massachusetts Bay from 1623–1636*. C. C. Little and J. Brown, Boston.

Zydlewski, Joseph, and Stephen D. McCormick. 1997. "The ontogeny of salinity tolerance in the American shad, *Alosa sapidissima*." *Canadian Journal of Fisheries and Aquatic Sciences* 54: 182–89.

Index

Barbara Brennessel was born in Brooklyn, N.Y. She earned a BS in biology from Fordham University and a PhD in biochemistry from Cornell University. After five years of postdoctoral work and teaching at a number of medical schools, she spent over thirty years at Wheaton College in Norton, Mass., where she taught courses such as Biochemistry, Microbiology and Immunology, and Nutrition. In addition to laboratory studies on population genetics of diamondback terrapins, spotted turtles, and oysters, she initiated field research projects that resulted in her previous books: *Diamonds in the Marsh: A Natural History of the Diamondback Terrapin* and *Good Tidings: The History and Ecology of Shellfish Farming in the Northeast.* She is Professor Emerita at Wheaton and lives in Wellfleet, Mass., where she continues to work on diamondback terrapin research and conservation, is a member of the Shellfish Advisory Board, and serves on the board of Friends of Herring River.